Richard Bowdler Sharpe

Monograph of the birds of paradise and Ptilonorhynchidae

Volume 2

Richard Bowdler Sharpe

Monograph of the birds of paradise and Ptilonorhynchidae
Volume 2

ISBN/EAN: 9783742867117

Manufactured in Europe, USA, Canada, Australia, Japa

Cover: Foto ©berggeist007 / pixelio.de

Manufactured and distributed by brebook publishing software
(www.brebook.com)

Richard Bowdler Sharpe

Monograph of the birds of paradise and Ptilonorhynchidae

MONOGRAPH

OF

THE PARADISEIDÆ,

OR

BIRDS OF PARADISE,

AND

PTILONORHYNCHIDÆ,

OR

BOWER-BIRDS.

BY

R. BOWDLER SHARPE, LL.D, F.L.S, F.Z.S, ETC.,

ASSISTANT KEEPER, ZOOLOGICAL DEPARTMENT, BRITISH MUSEUM; HOLDER OF THE GOLD MEDAL FOR SCIENCE FROM
H.I.M. THE EMPEROR OF AUSTRIA; HON. MEMBER OF THE NEW ZEALAND INSTITUTE;
FOREIGN MEMBER OF THE ROYAL ACADEMY OF SCIENCES AT LISBON;
HON. MEMBER OF THE IMPERIAL SOCIETY OF NATURALISTS OF MOSCOW; FOREIGN MEMBER OF THE ZOOLOGICAL SOCIETY OF FRANCE;
FOREIGN MEMBER OF THE AMERICAN ORNITHOLOGISTS' UNION; MEMBER OF THE HUNGARIAN ORNITHOLOGICAL CENTRAL BUREAU;
MEMBER OF THE BRITISH ORNITHOLOGISTS' UNION; HON. MEMBER OF THE MANCHESTER LITERARY
AND PHILOSOPHICAL SOCIETY, AND MEMBER OF THE MARLBOROUGH MICROSCOPIC AND NATURAL HISTORY SOCIETY;
AUTHOR OF A MONOGRAPH OF THE ALCEDINIDÆ, A MONOGRAPH OF THE HIRUNDINIDÆ (WITH CLAUDE W. WYATT);
AND OF VOLS. I, II, III, IV, VI, VII, X, XII, XIII, XVII AND XXIII, BRIT. MUS.
OF THE 'CATALOGUE OF BIRDS IN THE COLLECTION OF THE BRITISH MUSEUM,' ETC., ETC.

IN TWO VOLUMES.

VOLUME II.

LONDON:
HENRY SOTHERAN & CO.,
37 PICCADILLY, W. | 140 STRAND, W.C.
1891-1898.

LIST OF CONTENTS.

VOLUME II.

LIST OF PLATES.

VOLUME II

PHONYGAMA KERAUDRENI (*Less. & Garn.*).

Keraudren's Manucode.

Barita keraudreni, Less. & Garn. in Férussac, Bull. Sc. Nat. vii. p. 130 (1826).—Id. Voy. Coquille, Zool. Atlas, pl. 13 (1826).

Phonygama keraudreni, Less. Dict. Class. xiii. p. 309 (1828).—Id. Man. d'Orn. i. p. 141 (1828).—Id. Voy. Coquille, Zool. i. pt. 2. p. 636 (1828).—Id. Traité, p. 344 (1831).—Id. Compl. Buff., Ois. p. 403, cum tab. (1838).—Gray, Gen. B. ii. p. 303 (1845).—Bp. Consp. i. p. 358 (1850).—Rosenb. Nat. Tijdschr. Nederl. Ind. xxv. p. 236 (1863).—Id. J. f. O. 1864, p. 129.—Finsch, Neu-Guinea, p. 173 (1865).—Sharpe, Cat. Birds Brit. Mus. iii. p. 160 (1877).—Id. Journ. Linn. Soc. Zool. xvi. p. 442 (1882).—Kukenthal-Deckmann, Ann. Mus. Caen, i. p. 41 (1908).—D'Hamonv. Bull. Soc. Zool. France, 1896, p. 510.

Chalybæus cornutus, Cuvier, Règne Anim. i. p. 354 (1829).

Phonygama lessoni, Swains. Classif. B. ii. p. 264 (1837).

Manucodia keraudreni, Sclater, Journ. Linn. Soc. vi. p. 162 (1858).—Gray, Proc. Zool. Soc. 1858, p. 194.—Id. Cat. B. New Guinea, pp. 37, 59 (1859).—Id. Proc. Zool. Soc. 1859, p. 158.—Id. Proc. Zool. Soc. 1861, p. 430.—Id. Handl. B. iii. p. 17, no. 6238 (1870).—Elliot, Monogr. Parad. pl. 8 (1873).—Pavesi, Ann. Mus. Civic. Genov. vi. p. 315, tav. x. (1874).—Beccari, op. cit. vii. p. 713 (1875).—Salvad. op. cit. vii. p. 761 (1875), ix. pp. 48, 189 (1876), x. p. 150 (1877).—Pavesi, op. cit. ix. p. 64 (1876 cf?).—D'Albert. op. cit. x. pp. 13, 120 (1877).—Ramsay, Proc. Linn. Soc. N. S. W. iii. p. 102 (1876), iv. p. 97 (1879).—D'Albert. & Salvad. Ann. Mus. Civic. Genov. xiv. p. 95 (1879).—D'Albert. Nuova Guinea, pp. 502, 584, 588 (1880).—Salvad. Orn. Papuasia, etc. ii. p. 510 (1881).—Id. op. cit. iii. p. 551 (1882).—Musschenbr. Dagboek, pp. 195, 228 (1883).—Rosenb. Mitth. aus Ver. Wien, 1885, p. 40.

Chalybæus keraudreni, Schlegel, Mus. Pays-Bas, Coraces, p. 120 (1867).—Rosenb. Reis naar Zuidoostoröil, p. 47 (1867).—Id. Malay. Arch. pp. 320, 536 (1879).

Chalybæus keraudreni, Schlegel, Dierent. p. 175 (c. 1869).—Id. Nederl. Tijdschr. Dierk. iv. p. 50 (1871).

Phonygama jamesi, Sharpe, Cat. Birds Brit. Mus. iii. p. 191 (1877).—Id. Proc. Linn. Soc. xiii. p. 500 (1877).—Elliot, Ibis, 1878, p. 56.

Manucodia keraudreni (jamesi), Ramsay, Proc. Linn. Soc. N. S. W. iii. p. 265 (1879).

This Manucode is an inhabitant of New Guinea and the Aru Islands. In the former it is widely distributed, for it has been found by Lesson and Wallace near Dorey, and D'Albertis at Andai, by Beccari at Warbusi and Ramoi, and by the late Mr. Bruijn's hunters at Sorong, Nirbo, Mansinam, and in the Arfak Mountains.

D'Albertis met with the species on the Fly River, and again at Hall Bay and Yule Island, while Dr. E. P. Ramsay has recorded it from the vicinity of Port Moresby. At Aleya also the late Dr. James procured the specimen which I named after him, *Phonygama jamesi*. Count Salvadori considers that this species cannot be upheld, and I must confess that, with the series of specimens in the British Museum, I am at present unable to decide the question, though a re-examination of the type shows me that none of the other New Guinea birds have such a steely-green head.

The specimen procured by the late Carl Hunstein at East Cape also seems to me to resemble Dr. James's example. In the Aru Islands the present species has also been obtained by Von Rosenberg and Beccari; but, as the Count points out, there are some slight differences in specimens from these islands, the size being perceptibly larger, and the tint of the metallic gloss being of a more steel-blue character, as in examples from Sorong.

Nothing appears to have been written about the habits of this Manucode, beyond the statement of Von Rosenberg's that it was very rare in the Aru Islands, where it is only found in the woods, and not universally on the coast and in the interior forests like *M. atra*; its food consists of insects, small crustaceans, and fruit.

The peculiar trachea of these Manucodes is well known, and Professor Pavesi has written a memoir on the subject, founded on examples sent to him by Dr. Beccari. According to Pavesi the male has the trachea external and with several coils, the adult female has but a single coil, while the young females bear no trace of an external trachea at all.

The following description is copied from my third volume of the "Catalogue of Birds in the British Museum":—

Adult male. General colour above burnished steel-blue, the feathers of the head velvety in texture, as well as those of the sides of the face and chin; from the sides of the occiput extend two long tufts of steel-blue feathers; plumes of the hind-neck and throat also lanceolate; under surface of body burnished steel-blue, washed with green on the abdomen; wings and tail purple, the edges of the quills steel-blue, the latter black on their inner web; bill and legs black; iris red. Total length 13 inches, culmen 1·3, wing 6·35, tail 5·55, tarsus 1·25.

Adult female. Similar to the male, but a little smaller. The young bird, according to Count Salvadori, is blackish violet with scarcely any gloss.

The Plate represents an adult bird in two positions, drawn from a specimen in the British Museum.

[78]

PHONYGAMA PURPUREO-VIOLACEA, Meyer.

Purple-and-Violet Manucode.

Phonygama purpureo-violacea, Meyer, Zeitschr. ges. Orn. ii. p. 372, Taf. xv. (1885).—Id. Ibis, 1886, p. 262.—
D. Hamonn. Bull. Soc. Zool. France, xi. p. 419 (1886).—Sharpe in Gould's B. New Guin. vol. i. pl. 35
(1887).—Salvad. Agg. Orn. Papuasia, ii. p. 148 (1890).—De Vis, Ann. Rep. Brit. New Guin. p. 60
(1890).—Id. Colonial Papers, no. 163, p. 112 (1890).—Id. Ibis, 1891, p. 36.—Salvad. Agg. Orn.
Papuasia, iii. p. 239 (1891).

Writing in 1887, in the late Mr. Gould's 'Birds of New Guinea,' I suggested that the present species might possibly be the same as *Phonygama hunsteini*, described by me from a specimen procured by Mr. Hunstein at East Cape in South-eastern New Guinea. I came to this conclusion after an examination of a series of birds collected by Mr. H. O. Forbes in the Astrolabe Mountains. During the six years which have elapsed since I last examined into this question I have seen but few specimens of *P. purpureo-violacea*, but every one of them has so far confirmed the characters assigned by Dr. Meyer to the species. I now therefore consider that *P. hunsteini* must be kept distinct from the other species of the genus *Phonygama*, differing especially in its larger size.

I was at one time inclined to unite the present species with *P. jamesi*, and made the following remarks:—

"The series before us at the present moment leaves very little doubt that the *Phonygama* recently described by Dr. Meyer, and figured by us in the accompanying Plate, is distinct from *P. keraudreni* and *P. hunsteini*; but it is apparently the same as *Phonygama jamesi*, a species described by me in 1877 from Aleya, in South-eastern New Guinea. The chief difference between these two species is that *P. purpureo-violacea* is more purple above and steel-blue below, and *P. jamesi* is metallic green above and steel-green below. But between these extremes of colour every transition is found in the series now before us, and it should be noted that the type specimen of *P. jamesi* is moulting, and that the old feathers of the wing are very dull purple, while the new ones are bright purplish blue externally. In fine, without asserting dogmatically that *P. jamesi* and *P. purpureo-violacea* are the same, we have very little doubt in our own minds that they are, and that the steel-blue and green shades become gradually faded into purple or purplish blue."

On looking over the series of specimens in the British Museum, I am now inclined to think that *P. jamesi* may be distinct from *P. purpureo-violacea*, the latter being entirely purple above and below instead of steel-blue or steel-green.

The following description is taken from a specimen procured at 7000 feet elevation on the Owen-Stanley Range by Mr. Goldie and now in the British Museum:—

Adult male. General colour above purple, slightly shot with steel-blue or purplish blue on the lower back; wings rather more bronzy purple than the lower back, inclining to purplish blue on the margins of some of the feathers; the small coverts adjoining the bastard-wing steel-green; quills black, externally purple, the primaries shaded with steel-blue on the margins; tail-feathers purple, the inner webs black, the centre feathers somewhat bronzy; head with crested lateral tufts metallic steel-green all over, somewhat inclining to oily-green under certain lights, the hinder neck metallic purplish blue; sides of face and throat metallic steel-green, shading off on the fore neck and underparts into dark steel-blue, marked with purple; under wing-coverts like the breast; quills black below. Total length 10·5 inches, culmen 1·3, wing 6·4, tail 4·7, tarsus 1·4.

The figures in the Plate represent an adult bird in two positions, drawn from a specimen procured by Mr. Goldie in the Owen-Stanley Range and now in the British Museum.

PHONYGAMA HUNSTEINI, *Sharpe.*

PHONYGAMA HUNSTEINI, *Sharpe*.

Hunstein's Manucode.

Phonygama hunsteini, Sharpe, Journ. Linn. Soc. London, xvi. pp. 442, 443 (1882).—D'Hamonv. Bull. Soc. Zool. France, 1886, p. 510.—De Vis, Ann. Rep. Brit. New Guinea, p. 60 (1890).—Id. Colonial Papers, no. 103, p. 112 (1890).—Id. Ibis, 1891, p. 34.—Salvad. Agg. Orn. Papuasia, ii. p. 149 (1890), iii. App. p. 232 (1891).—Sharpe, Bull. Brit. Orn. Club, iv. p. xiii (1894).—Rothschild, Novit. Zool. iii. pp. 11, 233, 235 (1896).

Manucodia hunsteini, Salvad. Ann. Mus. Civic. Genov. xxxiv. p. 426 (1892).—Id. Orn. Papuasia, iii. App. p. 531 (1892).

Manucodia thomsoni, Tristram, Ibis, 1889, p. 554.

Phonygama thomsoni, Salvad. Agg. Orn. Papuasia, ii. p. 149 (1890), iii. App. p. 230 (1891).—Id. Ann. Mus. Civic. Genov. (2) x. p. 638 (1891).—De Vis, Rep. Coll. British New Guinea, p. 112 (1890).—Rothschild, Novit. Zool. iii. pp. 11, 235 (1896).

This Manucode was described by me in 1882 from a specimen sent by the late Carl Hunstein, and supposed to be from South-eastern New Guinea. That this locality is erroneous can hardly be doubted, as the real habitat of the species is now known to be Ferguson Island, in the D'Entrecasteaux group. Here it has been found by Mr. Basil Thomson, and also by Mr. Albert Meek, who says that it is met with in the hills, but seldom below 1500 feet.

Hunstein's Manucode is a large species, remarkable for its dull coloration, as compared with the metallic colours which adorn the other species of the genus. The green colour of the head, contrasting with the purplish-black colour of the upper and under parts, is also of a similar tint and of an oily-green lustre, without any remarkable gloss. Another peculiarity of the species, which both myself and Canon Tristram failed to observe, is the "hen"-shaped tail, as the Hon. Walter Rothschild points out. The tail-feathers slope downwards on each side from the central pair; and Mr. Rothschild, who has several specimens in his collection, remarks:—"The webs of the central rectrices in the fully adult male stand nearly perpendicular at the tip, but they are not twisted so far as to open again as they do in *Manucodia comrii*." The female does not differ in colour from the male, but the wing is a trifle shorter.

A young bird in Mr. Rothschild's museum is nearly black, with a slight gloss of dull steel-green, but with no purple on the back, wings, or tail. The head is black, with scarcely any steel-green gloss, and the crest-feathers are scarcely indicated. The under surface of the body is dull black, with scarcely any gloss except on throat.

Mr. Basil Thomson says that the convolutions of the windpipe are extraordinary, the latter being coiled under the skin.

The figure in the Plate has been drawn from the type-specimen in the British Museum.

PHONYGAMA JAMESI, *Sharpe.*

James's Manucode.

Manucodia keraudreni, Salvad. Ann. Mus. Civ. Genov. ix. p. 46 (1876, nec Less.).

Phonygama jamesi, Sharpe, Cat. Birds Brit. Mus. iii. p. 181 (1877).—Id Journ. Linn. Soc. xiii. p. 500 (1877).—Elliot, Ibis, 1878, p. 56.

Manucodia keraudreni, Salvad. Orn. Papuasia e delle Molucche, ii. p. 510 (1881, pt.).—Id. op. cit. iii. p. 551 (1882).

Phonygama keraudreni (nec Less.), Sharpe, Journ. Linn. Soc. xvi. p. 442 (1882).

I first described this species from a specimen procured at Aleya by the unfortunate Dr. James, who was murdered by the natives of South-eastern New Guinea.

Mr. D. G. Elliot, in 1878, expressed his opinion that *P. jamesi* could not be specifically separated from *P. keraudreni*, and Count Salvadori united it to the latter species. In deference to the opinion of these two excellent authorities on the *Paradiseidæ*, I myself acquiesced in the suppression of *P. jamesi* as a species, but not without a protest that I believed it to be really distinct. After a careful re-examination of the specimens in the British Museum and the Rothschild collection, I have come to the conclusion that *P. jamesi* is really distinct from *P. keraudreni*; and in this determination I am upheld by the Hon. Walter Rothschild, who has made a special study of the *Paradiseidæ*.

P. jamesi, though closely allied to *P. keraudreni*, is of a bright steel-green colour below, and has the throat, the sides of the head, the neck, and crest-plumes green instead of steelblue as in *P. keraudreni*, which also shows a sheen of purplish on these portions of the body.

The typical specimen of *P. jamesi* was procured by Dr. James at Aleya, on the mainland of South-eastern New Guinea, near Yule Island. A second specimen was procured by Mr. A. Goldie on the Laloke River, and the late Carl Hunstein also met with the species near East Cape. Both these specimens were recorded by me in the 'Journal of the Linnean Society' (*ll. cc.*), but they were not retained by the British Museum, which, however, possesses a second example, in addition to the type, from Mr. Broadbent's collection made in the interior of S.E. New Guinea, inland from Port Moresby.

The following description of the type specimen is taken from my third volume of the 'Catalogue of Birds':—

General colour above burnished green, with a slight shade of purplish-blue here and there on the back; under surface of body burnished green like the upper, with a subterminal lustre of purplish-blue on some of the feathers; head and neck all round of a burnished oily green, the plumes of the crown close-set and velvety, those of the neck and throat narrowly lanceolate; from the occiput two long tufts of green feathers; wings and tail purple, the wing-coverts burnished green like the back, the quills and tail-feathers black on their inner webs, except the innermost secondaries, which are entirely purple: bill and legs black. Total length 12·3 inches, culmen 1·3, wing 6·2, tail 4·9, tarsus 1·45.

The species is so closely allied to *P. keraudreni* that a separate figure has not been considered necessary.

PHONYGAMA GOULDI (Gray).

Gould's Manucode.

Manucodia keraudreni (nec Less.), Gould, Birds of Australia, Suppl. pl. 9 (1855).
Manucodia gouldi, Gray, Proc. Zool. Soc. 1859, p. 158, note.—Gould, Handb. B. Austr. i. p. 236 (1865).—Gray, Handl. B. ii. p. 17, no. 6250 (1870).—Masters, Proc. Linn. Soc. N. S. Wales, i. p. 50 (1877).
Phonygama gouldi, Sharpe, Cat. Birds Brit. Mus. iii. p. 161 (1877).—Finsch, Vög. der Südsee, p. 37 (1884).
Manucodia (*Phonygama*) *gouldi*, Ramsay, Tab. List Austr. Birds, p. 11 (1888).
Phonygama keraudreni (nec Less.), Witmer Stone, Proc. Acad. Nat. Sci. Philad. 1891, p. 446.

This species represents in the Cape York Peninsula of Australia the species of *Phonygama* which inhabit New Guinea, such as *P. keraudreni*, *P. jamesi*, *P. purpureo-violacea*, and *P. hunsteini*. In the occurrence of a true *Phonygama* in North-eastern Australia we may have a parallel case to the other Papuan forms which are met with on the Australian continent, viz. *Cnemorius*, *Ptilopterus*, *Craspedophora*, &c.

Gould's Manucode was discovered in the Cape York Peninsula by the late John Macgillivray during the voyage of the "Rattlesnake." For some time it was believed by Gould to be identical with *P. keraudreni* of New Guinea, but the differences were pointed out by Gray, and the species is now generally admitted to be distinct. It is a green bird and has the wings green, like the rest of the plumage, without any of the steel-blue or purple reflections seen in the New Guinea forms.

Dr. Otto Finsch met with the species near Somerset in the Cape York Peninsula, where he found it by no means common. It keeps out of sight, but its cry is often heard and resembles the bray of a child's toy-trumpet.

The following description is taken from my third volume of the 'Catalogue of Birds':—

General colour steel-green, of a somewhat oily-green cast on the rump and upper tail-coverts; wings coloured like the back, the coverts and the outer webs of the quills with a slight shade of purplish blue, this colour being also faintly indicated on the interscapulary region; wings black on the inner webs, excepting the innermost secondaries, which are bluish green; tail black, glossed with deep purplish blue, the feathers greenish on their outer edges under certain lights; head green, as also the two long tufts projecting from each side of the occiput; sides of face, throat, and underparts green, the abdomen with an oily-green shade, the feathers of the throat pointed and lanceolate, these parts somewhat shaded with steel-blue under certain lights; bill and legs black; "iris ochre-yellow" (O. Finsch). Total length 11·5 inches, culmen 1·25, wing 6·15, tail 5·15, tarsus 1·55.

It has not been considered necessary to give a separate figure of this species.

MANUCODIA CHALYBEATA. *Penn.*

MANUCODIA CHALYBEATA (*Pennant*).

Green Manucode.

Le Calybé de la Nouvelle Guinée, Daubent. Pl. Enl. iii. pl. 364.—Montbeill. Hist. Nat. Ois. iii. p. 208 (1775).
Oiseau de Paradis vert, Sonn. Voy. Nouv. Guin. p. 161, pl. 99 (1776).
Le Chalybé, Forst. Zool. Ind. p. 39 (1781).
Paradisea chalybeata, Penn. Faunula Indica, in Forst. Zool. Ind. p. 40 (1781).
Blue-green Paradise Bird, Lath. Gen. Syn. i. pt. 2, p. 482 (1782).
Manucodia chalybs, Bodd. Tabl. Pl. Enl. p. 39 (1783).—Sclater, Pr. Journ. Linn. Soc. ii. p. 162 (1858).—Gray, Cat. B. New Guin. p. 37, pl. (1859).—Elliot, Mon. Parad. pl. 6 (1873).—Salvad. Ann. Mus. Gen. vii. p. 781 (1875).—Peters, op. cit. ix. p. 34, fig. 2 (1876).—Gould, B. New Guin. i. pl. 34 (1877).—D'Hamonv. Bull. Soc. Zool. France, xi. p. 510 (1886).
Paradisea viridis, Scop. Del. Flor. et Faun. Insub. ii. p. 88 (1786).—Gm. Syst. Nat. i. p. 402 (1788).
Paradisea chalybea, Lath. Ind. Orn. i. p. 197 (1790).—Shaw, Gen. Zool. vii. p. 504, pl. 71 (1809).
Le Calibé, Audeb. et Vieill. Ois. Dor. ii. p. 24, pl. 10 (1802).—Levaill. Ois. de Paradis, p. 64, pl. 23 (1806).
Cracticus chalybeus, Vieill. Nouv. Dict. d'Hist. Nat. v. p. 355, pl. 330, fig. 1 (1816).—Id. Enc. Méth. p. 904 (1823).—Steph. in Shaw's Gen. Zool. xiv. pt. 2, p. 61 (1826).
Barita viridis, Cuv. Règn. An. i. p. 348 (1817).—Less. Man. i. p. 140 (1828).
Barita chalybea, Wagl. Syst. Av. *Paradisea*, sp. 2 (1827).—S. Müll. Verhandl. Land- en Volkenk. p. 22 (1839–44).—Temm. Pl. Col. Tab. Méth. p. 9 (1840).
Chalybaeus paradiseus, Cuv. Règn. An. i. p. 354 (1829).
Phonygama chalybaea, Less. Tr. d'Orn. p. 344 (1831).
Phonygama chalybea, Swains. Classif. B. ii. p. 264 (1837).—Finsch, Neu-Guinea, p. 172 (1865).
Phonygama viridis, Less. Compl. de Buff., Ois. p. 404 (1838).—Gray, Gen. B. ii. p. 363 (1846).—Bp. Consp. Av. i. p. 380 (1850).
Manucodia viridis, Gray, Gen. and Subgen. B. p. 65 (1855).—Id. P. Z. S. 1858, p. 194.—Id. Cat. B. New Guin. p. 38, pl. (1859).—Id. P. Z. S. 1861, p. 436.—Id. Hand-l. B. ii. p. 13, no. 6257, pl. (1870).—Scl. P. Z. S. 1876, p. 411.—Muschenbr. Dagboek, pp. 193, 228 (1883).—Rosenb. Mitth. orn. Ver. Wien, 1885, p. 40.
Chalybaeus viridis, Schl. Handl. Dierk. i. p. 334 (1857).—Id. Dieren. Vog. p. 175.—Id. Ned. Tijdschr. Dierk. iv. p. 49 (1871).
Chalybaeus viridis, Schl. Mus. P.-B., Coraces, p. 122 (1867).—Sunder. Meth. Av. Tent. p. 45 (1872).—Rosenb. Malay. Arch. pp. 356, 548 (1879).
Manucodia chalybeata, Salvad. Ann. Mus. Gen. vii. p. 968 (1875), xi. p. 389 (1878), x. p. 156 (1877).—D'Alb. et Salvad. op. cit. xiv. p. 94 (1879).—Meyer, Abhdl. Vog.-Skelet. p. 5, Taf. vii. et vii.a (1879).—Salvad. Orn. Papuasia, ii. pp. 498, 650 (1881).—Sharpe, Journ. Linn. Soc., Zool. xvi. p. 442 (1882).—Salvad. Orn. Papuasia, iii. p. 551 (1883).—Guillem. P. Z. S. 1885, p. 649.—Meyer, Zeitschr. ges. Orn. ii. p. 374 (1885).—Salvad. Ibis, 1886, p. 155.—Finsch & Meyer, t. c. pp. 241, 242.—Salvad. Agg. Orn. Pap. ii. p. 167 (1890).
Manucodia atra, D'Alb. (nec Less.) Syds. Mail, 1877, p. 248.—Id. Ann. Mus. Gen. x. p. 20 (1877).—Ramsa Pr. Linn. Soc. N. S. W. iii. p. 101, pl. ? (1877).
Manucodia chalybea, Sharpe, Cat. B. Brit. Mus. iii. p. 182 (1877).—Euden-Deslongch. Ann. Mus. d'Hist. Nat. Caen, i. p. 43 (1880).

CONSIDERING the number of years that this species has been known to science, it is certainly curious that so little has been recorded concerning its manners and habits; and, as will be seen below, even the exact characters of its plumage and its differences from allied species are still matters of some conjecture.

The Green Manucode is principally known from New Guinea, but it cannot be said to be as common as *M. atra* in any part of the great island. The most perfect specimens which I have yet seen was obtained by Mr. Wallace at Dorey, and beyond that example the British Museum does not possess a single authentic specimen from North-western New Guinea. The species has, however, been extensively procured by recent travellers in the Arfak Mountains, at Andei and Mansinam by Bruijn's hunters, and at Passi, Warmendi, and Profi by Dr. Beccari. D'Albertis met with it in Hatam, and Bruijn received specimens from Nappan, while Beccari records the species from Dorei-hum. The bird from Rubi, supposed at first to be the true *M. chalybeata*, has since been separated from it as *M. rubiensis* by Dr. A. B. Meyer.

Solomon Müller obtained the present species at Lobo, and D'Albertis met with it on the Fly River, which up to the present time has been supposed to be its most easterly range. I find, however, that since the publication of the 'Catalogue of Birds' the British Museum has received two specimens which mark a still more easterly extension of the known range of this Bird of Paradise in New Guinea. The late Hon. Hugh Romilly presented to the Museum some few years ago a valuable collection of birds from the Astrolabe Range, and amongst them I find a skin of the true *M. chalybeata*. A second specimen was presented by Sir James Ingham, the bird in question having been procured by his son in Cloudy Bay.

The statements as to the occurrence of this bird in Salawati, Waigiou, and the Aru Islands are apparently erroneous, as has been pointed out by Count Salvadori. It is found, however, on the island of Mysol, where the well-known Dutch traveller, Hoedt, met with it.

Dr. Guillemard, who obtained a male specimen in Mysol, has stated his opinion that *M. atra* and *M. chalybeata* are the same species and that the differences between them can be accounted for on the score of age. I confess that at one time I myself entertained the same idea, as there is so much variation in the plumage of *M. atra* that it seemed as if it must be a species of which *M. chalybeata* was the fully adult bird. As Dr. Guillemard points out, it seems improbable that there should be two such closely allied, yet distinct, species coexisting in the same districts; but, as Count Salvadori has said in his rejoinder, *M. atra* alone has been met with in the Aru Islands, and until the true *M. chalybeata* is found in the latter group it will be impossible to consider the two species identical. At present this statement is unanswerable.

The following description of Mr. Walter's Dorey specimen is copied from the 'Catalogue of Birds':—

Adult male. Head purple, the feathers compressed and close-set; the nape slightly washed with steel-greenish, as also the hinder neck and mantle; back rich purple, the feathers of the interscapulary region rather recurved; wings and tail rich purple, the inner webs of the feathers blackish, the outer wing-coverts somewhat shaded with steel-black; sides of the face and neck deep green, the feathers compressed and velvety like those of the crown; those of the chin, throat, and fore neck extending onto the sides of the neck, crinkled and curled and of an oily-green colour; the rest of the under surface deep purple, the feathers being tipped with this colour, less broadly on the vent and under tail-coverts, a few of the abdominal plumes with a slight greenish reflection; under wing-coverts black, the outer edge of the wing washed with green; bill and feet black; " iris red" (*Guillemard*). Total length 11·5 inches, culmen 1·65, wing 6·80, tail 5·9, tarsus 1·55.

The Plate is reproduced from Mr. Gould's 'Birds of New Guinea,' and represents an adult bird of the natural size.

MANCCODIA ATRA (Less)

MANUCODIA ATRA (Less.).

Glossy-mantled Manucode.

Barita viridis, var., Less. Man. d'Orn. i. p. 146 (1828).
Phonygama ater, Less. Voy. Coq., Zool. i. pt. 2, p. 638 (1828).—Id. Traité d'Orn. p. 344 (1831).—Id. Compl. Buff., Ois. p. 404 (1838).—Gray, Gen. B. ii. p. 303 (1846).
Phonygama atra, Bp. Consp. Av. i. p. 368 (1850).—Finsch, Neu-Guinea, p. 173 (1865).
Phonygama viridis (nec Scop.), Wall. Ann. Mag. Nat. Hist. (4) xx. p. 476 (1857).—Rosenb. Nat. Tijdschr. Nederl. Ind. xxv. p. 255 (1863).—Id. J. f. O. 1864, p. 171.
Manucodia atra, Scl. Journ. Linn. Soc. ii. p. 162 (1858).—Gray, P. Z. S. 1858, p. 194.—Id. Cat. B. New Guin. pp. 37, 59 (1859).—Id. P. Z. S. 1859, p. 158, 1861, p. 436.—Id. Handl. B. ii. p. 17, no. 6160 (1870).—Elliot, Monogr. Parad. pl. 3 (1873).—Salvad. Ann. Mus. Civic. Genov. vi. p. 781 (1875).—Id. & D'Albert. tom. cit. p. 828 (1875).—Pelz. Verh. zool.-bot. Gesellsch. Wien, 1875, p. 219.—Salvad. Ann. Mus. Civic. Genov. ix. pp. 40, 189 (1876), p. 156 (1877).—Sharpe, Journ. Linn. Soc. xiii. pp. 317, 500 (1877).—Id. Cat. Birds Brit. Mus. iii. p. 183 (1877).—Salvad. P. Z. S. 1878, p. 98.—Ramsay, Proc. Linn. Soc. N. S. Wales, iii. p. 101 (1878), p. 265 (1879), iv. p. 97 (1879).—D'Albert. Nuova Guinea, pp. 502, 584, 587 (1880).—Salvad. Orn. Papuasia e delle Molucche, ii. p. 504 (1881).—Id. Voy. 'Challenger,' ii. Birds, p. 82 (1881).—Forbes, P. Z. S. 1882, p. 347.—Gudea-Deslongch. Ann. Mus. Caen, i. p. 45 (1880).—Salvad. Orn. Papuasia, iii. App. p. 554 (1882).—Musschenbr. Dagboek, pp. 126, 229 (1883).—Ramsay, Proc. Linn. Soc. N. S. Wales, vio. p. 15 (1883).—Finsch, Vög. der Südsee, p. 29 (1884).—Meyer, Zeitschr. ges. Orn. i. p. 293 (1884).—Nehrk. J. f. O. 1885, p. 34.—Rosenb. Mitth. orn. Ver. Wien, 1885, p. 40.—Guillem. P. Z. S. 1885. p. 645.—Meyer, Zeitschr. ges. Orn. 1885, p. 374.—Id. Ibis, 1886, p. 342.—D'Hamonv. Bull. Soc. Zool. France, 1886, p. 210.—Salvad. Aggiunte Orn. Papuasia, ii. p. 148 (1890).—Id. Ann. Mus. Civic. Genov. (2) ix. p. 564 (1890).—Sharpe, Bull. Brit. Orn. Club, iv. p. xlv (1894).—Madarász, Aquila, i. p. 91 (1894).—Salvad. Ann. Mus. Civic. Genov. (2) xvi. p. 103 (1896).—Reichen. J. f. O. 1897, p. 272.
Manucodia torcha, pt., Gray, P. Z. S. 1858, p. 194, 1861, p. 436.—Id. Handl. B. ii. p. 17, no. 6157 (1870).
Manucodia chalybea, pt., Gray, Cat. B. New Guinea, p. 37 (1859).—Sclater, P. Z. S. 1881, p. 450.
Chalybeus ater, Schlegel, Mus. Pays-Bas, Coraces, p. 121 (1867).—Rosenb. Malay. Arch. pp. 370, 395, 556 (1879).
Chalybeus viridis (nec Scop.), Rosenb. Reis naar Zuidoostereil. p. 47 (1867).

In my third volume of the 'Catalogue of Birds in the British Museum' I separated *Manucodia atra* from *M. chalybeata*, on the strength of the different colour of the crinkled feathers of the throat and fore-neck, these being oily-green in the last-named species, and steel-black with an edging of velvety-black in *M. atra*. A far better definition of the distinctive characters of the two species is given by Count Salvadori in his 'Ornitologia della Papuasia,' where he separates *M. atra* from *M. chalybeata* on account of the smooth metallic feathering of the interscapulary region, whereas in *M. chalybeata* the feathers of this portion of the back have velvety-black transverse bands. I find that this difference holds good throughout the series in the British Museum, and even immature birds of both species can be recognized by these characters.

Manucodia atra has a somewhat extended distribution in the Papuan Sub-region, being found throughout the greater part of New Guinea, as well as on the adjacent Aru Islands, Mysol, Waigiou, Ghemien, Batanta, and Salawati. Its range in New Guinea includes the Arfak district in the north-west, as well as the Fly River and Port Moresby districts in the south, and it also extends to German New Guinea.

Besides being the most common species of the genus, it is found at a lower altitude than the others. Lesson states that its habits resemble those of Crows, and that it feeds on fruit on the large trees. Dr. A. R. Wallace, who observed it on the Aru Islands, writes:—"It is a very powerful and active bird; its legs are particularly strong, and it clings suspended to the smaller branches, while devouring the fruits on which alone it appears to feed." D'Albertis states that it lives on fruits, and especially on figs, while Von Rosenberg found it feeding on insects and worms. In South-eastern New Guinea the species is found in small troops and is very common in the neighbourhood of Port Moresby, and on the Laloke River, about twelve miles inland, according to Dr. E. P. Ramsay, who says that its cry has not the trumpet-like sound of *Phonygama keraudreni* and *P. gouldi*.

The difference in metallic tint, especially on the under surface of the body, is noteworthy in a series of specimens from different portions of the bird's range, some examples being much greener below than others, while the specimens from the Aru Islands are larger than those from New Guinea. Count Salvadori also remarks on the lighter coloration and larger size of the individuals obtained in Salawati and Batanta. Signor D'Albertis first described the trachea of this species as having only one coil, shaped like the letter S, resting in the depression of the furcula. The late Mr. W. A. Forbes confirmed D'Albertis's observations and described and figured the trachea in the 'Proceedings' of the Zoological Society for 1882. It is convoluted, but only to a small extent, merely forming a short loop lying on the inter-clavicular air-cell, between the rami of the furcula, much as in many specimens of the genus *Crax*. Mr. Forbes believed that in the female the trachea would be quite simple.

The following description is adapted from that given in my third volume of the 'Catalogue of Birds':—

Adult male. Head all round covered with compressed velvety plumes, steel-green without any purple reflections; the neck all round greenish like the head, the plumes slightly recurved and glistening with metallic ends; general colour of upper surface steel-black, shaded, according to the light, with metallic reflections of greenish or purple; wings and tail purple, with steel-black shades under certain lights, the quills externally glossed with greenish; under surface of body glossy steel-black with purplish or greenish reflections; under wing- and tail-coverts uniform with the breast; "bill and feet black; iris brilliant red" (*F. H. H. Guillemard*). Total length 16 inches, culmen 1·6, wing 7·2, tail 6·9, tarsus 1·5.

Adult female. Similar to the male, but smaller; "iris dull orange" (*F. H. H. Guillemard*). Total length 16 inches, culmen 0·45, wing 7, tail 6·8, tarsus 1·7.

The figure in the Plate represents an adult male, drawn from a Dorei specimen in the British Museum.

MANUCODIA COMRII, *Sclater.*

MANUCODIA COMRII, *Sclater*.

Curl-crested Manucode.

Manucodia comrii, Sclater, Proc. Zool. Soc. 1876, p. 459, pl. xlii.—Id. Ibis, 1876, p. 364.—Salvad. Ann. Mus. Gen.
ix. p. 191 (1876).—Sclater, P. Z. S. 1877, p. 43.—Gould, B. New Guin. i. pl. 33 (1877).—Ramsay, Proc.
Linn. Soc. N. S. W. iv. p 469 (1879).—Eudes-Deslongch. Ann. Mus. d'Hist. Nat. Caen, i. p. 47 (1880).—
Salvad. Orn. Papuasia, ii. p. 491 (1881).—Sharpe, Journ. Linn. Soc. Zool. xvi. p. 442 (1882).—
Salvad. Orn. Papuasia, iii. p. 551 (1882).—Musschenbr. Dagboek, pp. 195, 230 (1883).—Rosenb.
MT. orn. Ver. Wien, 1885, p. 49.—D'Hamonv. Bull. Soc. Zool. France, xi. p. 510 (1886).—Tristr.
Ibis, 1889, p. 554.—Salvad. Agg. Orn. Papuasia, ii. p 147 (1890).—Id. Ann. Mus. Gen. (2) x. p. 233
(1891).—Id Agg. Orn. Papuasia, iii. p. 230 (1891).—North, Rec. Austr. Mus. ii. p. 32, pl. vii. (1892).—
Meyer, Abhandl. k. zool. Mus. Dresden, 1892-93, no. 3, p 13 (1893)

This is the largest species of the Manucodes, and was discovered by Dr. Comrie in Huon Gulf, in South-eastern New Guinea, during the cruise of H.M.S. 'Basilisk.' Mr. Geisler also saw the species in this locality, but did not manage to procure a specimen. Dr. Meyer suggests that a comparison between the birds from New Guinea and D'Entrecasteaux Island should be instituted. In the British Museum are specimens from both the above-named localities, as well as from Normanby Island, and I cannot perceive the least difference between them, in size or colour. Mr. A. Goldie procured several specimens on Fergusson or D'Entre-casteaux Island, and the species has been met with in the same locality by Dr. Lewis, who also procured it on Goodenough Island.

The following is Dr. Sclater's original account of the species :—

"Dr. Comrie has placed in my hands for determination some bird-skins collected by him while serving as medical officer in H.M.S. 'Basilisk' during its recent survey of the south-east coast of New Guinea under the command of Captain Moresby. The collection contains thirteen skins, belonging to eleven species, of which one is quite new to science, and two others are only known from single specimens."

"This *Manucodia*," Dr. Sclater continues, " may be regarded as by far the finest and largest species of the genus yet discovered. It is immediately distinguishable from *M. chalybeia* and *M. atra* by its much larger size and longer bill, which is deeply sulcated at the nostrils. The characteristic curling of the feathers is extended to a greater degree, and pervades the whole of the head and neck. The feathers of the abdomen are black at the base, broadly margined with purple. Dr. Comrie obtained a single specimen of this fine bird in May 1874 in Huon Gulf. It was shot flying amongst the trees in the scrubby forest, about a quarter of a mile from the coast."

The egg of this species is described and figured by Mr. A. J. North in the 'Records' of the Australian Museum for 1892. The photographic illustration which accompanies Mr. North's paper represents an egg of the type of those of the Birds of Paradise with which we have recently been made familiar.

Mr. North gives the following account of this interesting discovery :—" The Trustees of the Australian Museum have lately received from the Rev. R. H. Rickard the egg of *Manucodia comrii*, taken by him on Fergusson Island, off the south-east coast of New Guinea, in July 1891. The Rev. Mr. Rickard informs me that from the 20th of June to the 20th of July he had at various times engaged, in company with his black boy, in shooting Manucodes on this island, but rarely saw a female bird. Early in July he found a nest of this species in the lower branches of a bread-fruit tree at a height of twenty-five feet from the ground. The female was on the nest, which was an open, loosely-made structure of vinelets and twigs, placed at the extremity of the branch. Having secured the bird, he found that she was in very indifferent plumage, as though she had been sitting for a long time, and the eggs, two in number, were chipped and just upon the point of hatching. The egg is an elongate ovoid in form, and is of a warm isabelline ground-colour, with purplish dots, blotches, and bold longitudinal streaks, uniformly dispersed over the surface of the shell, intermingled with similar super-imposed markings of purplish grey. Length 1·60 inch × 1·13 inch."

The following is the description of the type specimen, from Huon Gulf :—

Adult. Above velvety black, the feathers slightly recurved at the ends, with a subterminal mark of metallic green ; scapulars metallic steel-green, edged with velvety black : lesser and median wing-coverts steel-green, shot with purple, with a narrow fringe of velvety black ; greater coverts, primary-coverts, and quills metallic purple, shot with steel-blue at the end of the secondaries and especially on the innermost of these quills; primaries black, glossed with steel-blue near the ends and slightly glossed with purple externally; tail-feathers metallic purple, the two centre ones recurved and twisted over, with the webs decomposed; crown of head and neck covered with frizzled plumes; the lores, sides of face, and cheeks velvety black, with a green gloss under certain lights; the crown glossed with purplish and steel-blue and formed into a ridge over the eye; hind neck, sides of neck, and throat metallic oily-green, the feathers very close set and velvety to the touch; the fore neck and chest with longer and more thick-set plumes of green, somewhat steel-green in their gloss, and glossed with purple on the chest; remainder of under surface of body bronzy purple, with a distinct subterminal bar of velvety black on each feather; lower abdomen and thighs glossy steel-green, as also the under wing-coverts; under tail-coverts purple; bill and feet black. Total length 17·5 inches, culmen 2·4, wing 9·5, tail 7, tarsus 2·4.

The Plate represents an adult bird of the full size, drawn from the original specimen, now in the British Museum.

MANUCODIA ORIENTALIS, *Salvad.*

Eastern Manucode.

Manucodia chalybeata, part, Finsch & Meyer, Zeitschr. ges. Orn. 1895, p. 374.—Id. Ibis, 1886, p. 241.—
Salvadori, Aggiunte Orn. Papuasia e delle Molucche, ii. p. 147 (1890).—Id. Ann. Mus. Civic. Genov.
(2) x. p. 821 (1891).—Meyer, Abhandl. k. Mus. Dresd. 1890-91, no. 4, p. 32 (1891).—Id. J. f. O.
1892, p. 260.
Manucodia orientalis, Salvad. Ann. Mus. Civic. Genov. (2) xvi. p. 103 (1896).—Rothschild, Novit. Zool. iii.
p. 232 (1896).

According to Count Salvadori, this species differs from the true *M. chalybeata* in having a more slender bill, and the feathers of the neck have a more pronounced blue gloss, while the feathers above the eye are longer in the adult bird and form two ridges.

This form, which Mr. Rothschild regards merely as a race of *M. chalybeata*, inhabits Eastern New Guinea. It has been found on the coasts near Milne Bay and to the north of Huon Gulf, while Dr. Loria has procured it in several places on the Owen Stanley Mountains, where also Mr. Alfred Meek met with it. The adult specimens from South-eastern New Guinea in the British Museum differ from *M. chalybeata* from Dorey in having more steel-blue or green on the crown, the latter being decidedly purple in the Dorey bird.

According to Dr. Loria, the bill and feet and the iris are dull red. The bird feeds on fruit.

Of this and the following species it has not been considered necessary to give figures.

MANUCODIA JOBIENSIS, *Salvad.*

Jobi Manucode.

Manucodia jobiensis, Salvad. Ann. Mus. Civic. Genov. vii. p. 969 (1875), viii. p. 404 (1876), ix. p. 189 (1877).
—Sharpe, Cat. B. Brit. Mus. iii. p. 194 (1877).—Elliot, Ibis, 1878, p. 56.—Meyer, Abhidl. Vog.-Skel.
p. 56, Taf. vii. a (1879).—Essex-Denborugh, Ann. Mus. Caen, i. p. 45 (1880).—Salvad. Orn. Papuasia
e delle Molucche, ii. p. 392 (1881).—Musschenbr. Dagboek, pp. 196, 222 (1883).—Rosenb. Mitth.
a. n. Ver. Wien, 1885, p. 40.—Guillem. P. Z. S. 1885, p. 649.—Meyer, Zeitschr. ges. Orn. 1885,
p. 374.—Id. Ibis, 1886, p. 242.—D'Hamonv. Bull. Soc. Zool. France, 1886, p. 510.—Salvad. Aggiunte
Orn. Papuasia, ii. p. 147 (1890).—Sharpe, Bull. Brit. Orn. Club, iv. p. xvi (1894).

The Manucode of Jobi Island has been separated by Count Salvadori on the following characters :— The head is of a metallic green, instead of steel-blue as in *M. chalybeata*. The underparts are shining green, and the fore-neck and upper breast are green, with velvety black transverse bands, while the lower breast and abdomen are of a less lustrous green without the black velvety bands. In *M. chalybeata* the fore-neck and upper breast are glossy green with golden spots, and the rest of the under surface is steel-blue, inclining to violet under certain lights, and the feathers of the lower breast have a velvety black transverse band. The interscapular feathers incline more to green, and the black velvety edgings are less conspicuous.

The bill is more compressed, and commences at a more acute angle in the middle of the forehead, a difference which is well marked and constant. In the shape of the bill *M. jobiensis* approaches *M. atra*, which always has the culmen broader. In the above-mentioned characters *M. jobiensis* is intermediate between *M. chalybeata* and *M. atra*. No specimen of *M. jobiensis* is in the British Museum, but I notice considerable variation in the tints of the metallic lustre in a series of *M. chalybeata*.

Count Salvadori states that Dr. Beccari obtained the species at Sorie in Jobi, and Mr. Bruijn's hunters at Wonapi. Dr. Guillemard met with it near Ansus in November, and considered the species to be uncommon in the island. The bill and feet were black in the specimen he obtained, and the iris red. His measurements are less than those given by Count Salvadori, but, as Dr. Meyer remarks, these are probably attributable to a difference in sex.

The trachea is like that of *M. chalybeata* and the convolution is confined to a single loop. It is figured by Dr. Meyer (*l. c.*). In the female and young male of *M. chalybeata* the convolution of the trachea is not present.

MANUCODIA RUBIENSIS, Meyer.

Rubi Manucode.

Manucodia rubiensis, Meyer, Zeitschr. ges. Orn. ii. p. 374 (1885), iii. p. 36 (1886).—Id. Ibis, 1886, p. 242.—
    D'Hamonv. Bull. Soc. Zool. France, 1886, p. 510.—Salvad. Aggiunte Orn. Papuasia, ii. p. 147
    (1890).—Sharpe, Bull. Brit. Orn. Club, iv. p. xiii (1894).

_____

This species, of which I have not seen a specimen, has been described by Dr. A. B. Meyer. It is said to be similar to *M. chalybeata*, but is much smaller. Dr. Meyer writes:—" From Rubi, the most southern point of the Bay of Geelvink, which is very remarkable in its ornithology, two examples lie before me, which are distinguished by their small size from those of other localities. Moreover, the under surface of the neck seems to be green instead of blue, and the curling of the feathers is only very slightly developed. The bill is but little stronger than in *Phonygama keraudreni*."

Laglaize has also procured this species at Kafu.

LYCOCORAX PYRRHOPTERUS, Temm.

W. Hart del. et lith.

Walter, Imp.

LYCOCORAX PYRRHOPTERUS (Bp.).

Brown-winged Paradise-Crow.

Corvus pyrrhopterus, Bp. Consp. Av. i. p. 384 (1850, ex Temm. MSS. in Mus. Lugd.).

Lycocorax pyrrhopterus, Bp. Comptes Rendus, xxxvi. p. 829 (1853).—Id. Notes Coll. Delattre, p. 7, note (1854).—
Gray, P. Z. S. 1860, p. 365.—Finsch, Neu-Guinea, p. 173 (1866).—Berust. Nederl. Tijdschr. Dierk. ii.
p. 372 (1865).—Schl. op. cit. iii. p. 191 (1866).—Id. Mus. Pays-Bas, Coraces, p. 131 (1867).—Salvad.
Ann. Mus. Civ. Genov. vii. p. 781 (1875).—Sharpe, Cat. B. Brit. Mus. x. p. 165 (1877).—Salvad. Ann.
Mus. Civ. Genov. xvi. p. 198 (1880).—Eudes-Deslongchamps, Ann. Mus. d'Hist. Nat. Caen, i. p. 46
(1880).—Salvad. Orn. Papuasia, ii. p. 494 (1881).—D'Hamonv. Bull. Soc. Zool. France, 1886, p. 510.—
Salvad. Orn. Papuasia, Agg. ii. p. 146 (1890).

Pica pyrrhoptera, Schl. Bijdr. tot de Dierk. fol. pt. viii. p. 1, pl. 1 (1866).

Manucodia pyrrhoptera, Gray, Handl. B. ii. p. 17, no. 6261 (1870).—Meyer, Dagboek, pp. 197, 230 (1882).—
Rosenb. M. T. orn. Ver. Wien, 1886, p. 40.

THE three species of *Lycocorax*, which are structurally Birds of Paradise, so closely resemble the ordinary
Crows in appearance that they may naturally be placed at the end of the family *Paradiseidæ*, the members of
which are remarkable for the brilliancy and fantastic arrangement of their colouring. In this position they
are placed by Count Salvadori in his 'Ornitologia della Papuasia.'

The present species is apparently confined to the islands of Batchian and Halmahéra, and it is easily
recognised by the colour of the quills, which is reddish on the outer aspect of the primaries, the secondaries
being brown. The general colour of the plumage is black, with very slight greenish reflections under
certain lights.

Nothing seems as yet to have been recorded regarding the habits of this species.

Adult female. General colour above and below black, with a slight wash of dull oily green; tail black,
with an almost imperceptible wash of green on the outer web of some of the feathers; least wing-coverts
resembling the back; the rest of the wing brown, becoming paler and more reddish on the outer primaries,
the lower surface of the quills inclining to ashy brown; bill and feet black.

Total length 15·5 inches, culmen 1·75, wing 7·1, tail 6·1, tarsus 1·75.

According to Count Salvadori, there is no apparent difference in the colour of the sexes.

The description and figure are taken from specimens in the British Museum. The latter represents an
adult bird of the natural size.

LYCOCORAX OBIENSIS, *Bernst.*

LYCOCORAX OBIENSIS, *Bernst.*

Obi Paradise-Crow.

Lycocorax obiensis, Bernst. Journ. für Orn. 1864, p. 410.—Id. Nederl. Tijdschr. Dierk. ii. p. 350 (1865).—Schl. op.
cit. iii. p. 192 (1866).—Id. Mus. Pays-Bas, Coraces, p. 132 (1867).—Sharpe, Cat. Birds in Brit. Mus. iii.
p. 185 (1877).—Eudes-Deslongchamps, Ann. Mus. d'Hist. Nat. Caen, i. p. 47 (1880).—Salvad. Ann. Mus.
Civic. Genov. xvi. p. 199 (1880).—Id. Orn. Papuasia e delle Molucche, ii. p. 495 (1881).—Guillemard,
Proc. Zool. Soc. 1885, p. 553.—D'Hamonv. Bull. Soc. Zool. France, i. p. 510 (1896).—Sharpe in Gould's
B. New Guinea, i. pl. 36 (1888).—Salvad. Agg. Orn. Papuasia, ii. p. 146 (1890).
Manucodia obiensis, Gray, Hand-l. B. ii. p. 17, no. 6263 (1870).—Meyer-Schön. Dagboek, pp. 193, 236 (1883).—
Reichb. Mitth. orn. Ver. Wien, 1885, p. 40.

THE present species occurs in the islands of Obi Major and Obi Latitoo, in the Moluccas, where it was
discovered by the late Dr. Bernstein, who described the species. It has a distinct greenish wash on the
upper surface, and this character distinguishes it from its two allies *Lycocorax pyrrhopterus*, from Batchian
and Gilolo, and *L. morotensis*, from Morotai or Morty Island. The black secondary quills also distinguishes the species from *L. pyrrhopterus*, which has brown secondaries, and the specimen in the British
Museum justified me in separating it from *L. morotensis*, on account of the white bases to the quills of the
latter. Count Salvadori likewise adopted this difference, these characters of the wing, to the separation of
the two species; but when Dr. Guillemard visited the Obi Islands in 1884, he obtained five specimens of
L. obiensis, and of these the four males had the inner web of the primaries white at the base, so
that this distinction from *L. morotensis* is not upheld.

Dr. Bernstein states that he met with the species both in Obi Major and Obi Latitoo, but he found
it, like *L. morotensis*, a difficult bird to procure, as its home is in the thick forest. He describes the
note as resembling the word "whaak."

The following description of an adult bird is copied from the British Museum 'Catalogue of Birds,'
and is taken from a specimen in that institution:—

"General colour above and below of a dull rifle-green, somewhat glistening; tail black, the feathers
slightly washed with green on the outer web; quills blackish brown, paler towards the base on the inner web,
the least wing-coverts edged with dull green like the scapulars, the rest of the coverts and secondaries
slightly washed with green on the outer web, the primaries much paler brown; 'bill and feet black;
iris crimson' (*Guillemard*). Total length 13·5 inches, culmen 1·95, wing 7·75, tail 6·75, tarsus 1·9."

A female bird obtained by Dr. Guillemard was probably immature, as it was duller in colour and had the
wings lighter brown, the primaries buff, and the iris brown instead of crimson.

The Plate is reproduced from Mr. Gould's 'Birds of New Guinea,' and the figure represents an adult
bird of about the size of life, drawn from one of Dr. Guillemard's specimens.

LYCOCORAX MOROTENSIS, *Bernstein.*

Morty Island Paradise-Crow.

Lycocorax morotensis, Bernst. MSS.—Schlegel, Ibis, 1863, p. 119.—Bernst. J. f. O. 1864, p. 408.—Finsch, Neu-
Guinea, p. 173 (1865).—Schlegel, Ned. Tijdschr. Dierk. iii. p. 191 (1866).—Id. Mus. Pays-Bas, Coraces,
p. 132 (1867).—Sharpe, Cat. Birds Brit. Mus. iii. p. 186 (1877).—Salvad. Ann. Mus. Civ. Genov. xri
p. 309 (1880).—Id. Orn. Papuasia, ii. p. 456 (1881).—Eudes-Deslongch. Ann. Mus. d'Hist. Nat. Caen, i.
p. 47 (1880).—D'Hamonv. Bull. Soc. Zool. France, 1885, p. 510.—Salvad. Agg. Orn. Papuasia, ii.
p. 147 (1890).—Sharpe, Bull. Brit. Orn. Club, iv. p. xiv (1894).
Manucodia morotensis, Gray, Hand-l. B. ii. p. 17, no. 6263 (1870).
Manucodia morotensis, Musschenbr. Dagboek, pp. 197, 230 (1883).
Manucodia morotensa, Rosenb. Mitth. orn. Ver. Wien, 1885, p. 40.

THE present species represents in the island of Morotai, or Morty Island, the *Lycocorax pyrrhopterus* of
Batchian and Gilolo, and the *L. obiensis* of the Obi group. The restriction of this peculiar genus to these
small groups of islands in the Moluccas is one of the interesting facts connected with the distribution of the
Birds of Paradise.

L. morotensis is also found in the small island of Rau. It is easily distinguished from its two allies by the
white base to the inner web of the primaries.

The species was discovered in Morotai by the late Dr. Bernstein, who gave the following account of it :—
" Like the other species of the genus, it inhabits the thick woods and is rarely seen outside of them. It
generally lives in trees of moderate height, especially where they stand close together, in the tops of which
it hides closely, so that, though often heard, it is a very difficult bird to see. It is most easily observed, if
the hunter places himself in the early morning near some tree on the fruit of which the bird comes to
feed. But even then the greatest attention must be maintained, as the bird does not come flying in like a
Pigeon, but glides quietly from the top of one tree to the summit of another, lights for an instant on some
fruit-bearing bough, is seen for a second on the outer branches, and then dives into the thickest of the
foliage. In all its ways of life there is very little Crow-like, and it seems to feed exclusively on the fruit
of trees. Its cry is a short, interrupted, monosyllabic 'wukh' or 'wakh,' which is especially heard in the
morning and evening. My hunters fancied that the note had some similarity to the ringing bark of a dog,
and called the bird 'Burong andjing,' or 'Dog-bird.'"

Adult. General colour black above and below, the wings brown, the primaries lighter than the secondaries,
and having the base of the inner web conspicuously white ; tail black, shaded with dull green on the outer
web ; bill and feet black. Total length 17 inches, culmen 2, wing 8·4, tail 7, tarsus 2.

The description is taken from a specimen in the British Museum, and the figure in the Plate has been
drawn from the same bird.

PARADIGALLA CARUNCULATA, *Less.*

PARADIGALLA CARUNCULATA, Less.

Wattled Bird of Paradise.

Paradigalla carunculata, Less. Ois. Parad. p. 242 (1835).—Id. Rev. Zool. 1840, p. 1.—Bp. Consp. i. p. 414 (1850).—
Sclater, Proc. Zool. Soc. 1857, p. 6.—Id. Proc. Linn. Soc. ii. p. 164 (1858).—Wallace, Proc. Zool. Soc.
1862, p. 160.—Rosenb. Journ. für Orn. 1864, p. 133.—Wallace, Malay Arch. ii. pp. 257, 258 (1869).—
Elliot, Monogr. Parad. pl. 17 (1873).—Beccari, Ann. Mus. Civic. Genov. vii. p. 711 (1875).—Salvad. tom.
cit. pp. 794, 899, xi. p. 190 (1876).—Sclater, Ibis, 1876, p. 250.—Sharpe, Cat. Birds in Brit. Mus. iii.
p. 166 (1877).—Gould, Birds of New Guinea, i. p. 16 (1868).—Eudes-Deslongchamps, Ann. Mus. d'Hist.
Nat. Caen, i. p. 20 (1880) = Salvad. Orn. della Papuasia, etc. ii. p. 430 (1881).—Guillemard, Proc.
Zool. Soc. 1885, p. 651.—D'Hamonv. Bull. Soc. Zool. France, xi. p. 369 (1886).—Salvad. Agg. Orn.
Pap. ii. p. 151 (1890).—Wallace, Malay Arch. 2nd ed. pp. 455, 457 (1890).

Astrapia carunculata, Eydoux et Souleyet, Voy. 'Bonite,' Zool. i. p. 83, Atlas, Ois. pl. 4 (1841).—Gray, Gen. B. ii.
p. 326 (1846).—Id. Proc. Zool. Soc. 1858, p. 184. Id. List B. New Guinea, pp. 37, 39 (1859).—Id.
Proc. Zool. Soc. 1861, p. 436.—Finsch, Neu-Guinea, p. 179 (1865).—Rosenb. Malay Arch. p. 556
(1879).—Musschenbr. Dagboek, pp. 194, 227 (1883).—Rosenb. Mitth. orn. Ver. Wien, 1885, p. 40.

Paradisea carunculata, Schlegel, Journ. für Orn. 1861, p. 386.

At first sight there is nothing very attractive in the appearance of this Bird of Paradise, which might be
considered more curious than striking to look at. On a closer examination, however, it will be found that
it is clothed in velvety plumage of a beautiful texture, while its wattles, of three colours, are unique among
the family of Paradise-birds. In my 'Catalogue of Birds' I have placed the genus *Paradigalla* in close
proximity to *Astrapia*, and in this arrangement Count Salvadori concurs in the main, but he also points out
that in many of its characters it also approaches *Parotia* and *Lophorhina*. It has the velvety plumage of both
the latter genera, and has the first two primaries pointed, as in *Lophorhina*. In *Parotia* these quills are
curiously notched at the ends, and the secondaries are as long as the primaries, while the tail is much
graduated, the centre feathers being the longest. In all these characters *Paradigalla* assimilates to *Parotia*,
but it has the two centre tail-feathers very much lengthened and pointed, considerably exceeding the other
tail-feathers in length.

The early history of the species has been given in detail by Count Salvadori. It was first named by
Lesson in 1835, in the Synonymic Index to his 'Histoire Naturelle des Oiseaux de Paradis;' but the
description is not very complete, and he does not say whence he described the specimen or in whose
collection the bird was. It may probably have been the one mentioned by him in 1840 as being in the
collection of Dr. Abeille of Bordeaux. In 1841 Messrs. Eydoux and Souleyet, the Naturalists attached
to the 'Bonite,' described the present bird as *Paradigalla carunculata*, but made no allusion to Lesson's
having named the species. During the voyage of the 'Bonite' they procured two mutilated specimens in
New Guinea, one of which appears to have ultimately gone to Philadelphia. For many years these
imperfect skins remained the only examples known in Museums, and even Baron von Rosenberg did not
succeed in obtaining the species in perfect condition. The first examples of complete skins of the
Paradigalla were obtained by Dr. A. B. Meyer, and during recent years many have been procured by
Dr. Beccari and Mr. Bruijn's hunters.

Dr. Beccari has given the following note on his experience of the present species:—

"As to *Paradigalla*, I shot one from my hut, whilst it was eating the small fleshy fruits of an *Urtica*.
It likes to sit on the tops of dead and leafless trees, like the *Mino dumonti*. The finest ornaments of this bird
are the wattles, which in the dried skin lose all their beauty. The upper ones, which are attached one on
each side of the forehead, are of a yellowish-green colour; those at the base of the lower mandible
are blue, and have a small patch of orange-red beneath. The Arfaks call the *Paradigalla* 'Hoppoa.'"

Dr. Guillemard states that M. Laglaize told him that the colour of the caruncles was as follows:—
The upper caruncle is orange, the middle one bright leaf-green, and the lower one red. It will be noticed
that these colours are somewhat different from those given by Dr. Beccari; but as the latter
gentleman made his observations on specimens killed by himself, there can be no doubt as to their
accuracy.

The following description of the male is copied from my 'Catalogue of Birds':—

Adult male. General colour velvety black above and below, a little browner on the under surface; wings and tail black, the inner secondaries with a purplish gloss under certain lights; head glossed with metallic steel-green; forehead, lores, and base of lower mandible bare; over each nostril a small tuft of black feathers; on each side of the base of the bill an erect wattled skin; round the eye a ring of black plumes; space below and behind the eye bare: bill and legs black; "iris red" (*Guillemard*): "upper wattles, which are attached one on each side of the forehead, of a yellowish-green colour; those at the base of the lower mandible blue, having a small patch of orange underneath" (*Meccau*). Total length 11·2 inches, culmen 0·55, wing 6·15, tail 4·85, tarsus 1·9.

Adult female. Similar to the male, but smaller. Count Salvadori says that it is deep black, with an appearance of dull green on the crown and occiput, the wings and tail with scarcely any velvety appearance, and with a dull purplish shade under certain lights.

The figures in the Plate represent adult birds of the size of life. They are reproduced from Mr. Gould's 'Birds of New Guinea.'

MACGREGORIA PULCHRA, De Vis.

MACGREGORIA PULCHRA, De Vis.

Orange-wattled Bird of Paradise.

Maria macgregoriæ, Gigl. Boll. Soc. Geogr. Ital. p. 26 (1897 . descr. nulla).
Macgregoria pulchra, De Vis, Ibis, 1897, p. 251, pl. vii.

This species was discovered by Sir William MacGregor on Mount Scratchley, in British New Guinea, at a height of about 12,000 feet. Mr. De Vis writes :—" Three examples of this bird, all (presumably) males, were obtained by Sir W. MacGregor, in May 1896, during his journey across British New Guinea from Mambare to the Vanapa River. Mr. A. Giulianetti, his Excellency's collector, notes that ' the species is pretty common all over the Scratchley Range up to about 12,000 feet elevation.' " Mr. De Vis has forwarded to England one of the specimens procured by Sir W. MacGregor during the above-mentioned expedition. This specimen has been figured in the ' Ibis' for 1897, and has been presented to the British Museum. Dr. Sclater has called attention to the fact that Professor Giglioli (*l. c.*) is responsible for the publication of the name of *Maria macgregoriæ*—a name which appears to have been mentioned in a private letter addressed to Professor Giglioli by Sir W. MacGregor. As Dr. Sclater remarks : " Unfortunately the generic term ' *Maria* ' has been already employed in Zoology (Bigot, Rev. et Mag. de Zool. 1859, p. 311 : Diptera)."

The genus *Macgregoria* is undoubtedly closely allied to *Paradigalla*, but the different position of the fleshy wattles on the sides of the face renders farther comparison unnecessary.

Adult male. General colour above and below black, with scarcely any purplish gloss, and with a large bare orange wattle covering the whole of the ear-coverts and the region of the eye ; primaries orange-buff with black tips : iris red. Total length 13 inches, culmen 1·3, wing 7·4, tail 5·6, tarsus 2·4.

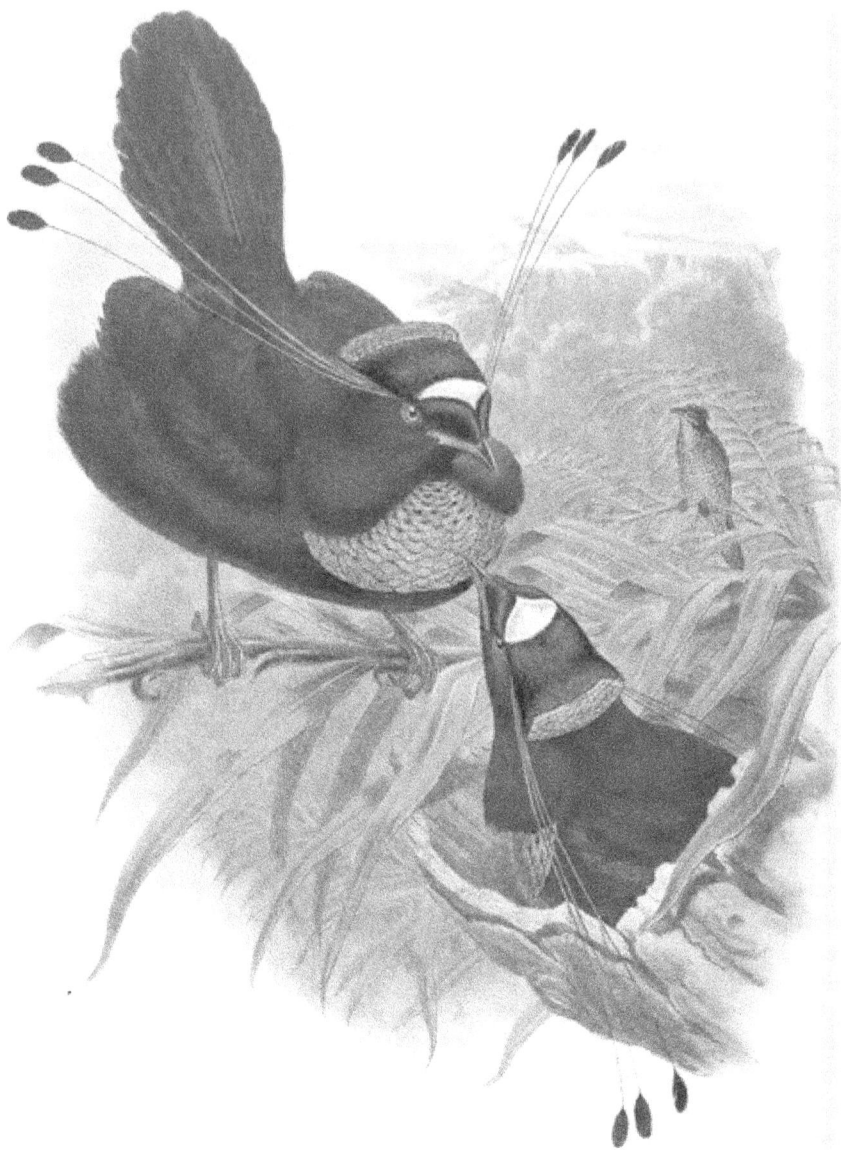

PAROTIA SEXPENNIS (*Bodd.*).

Six-plumed Bird of Paradise.

Le Sifilet de la Nouvelle Guinée, D'Aubent. Pl. Enl. pl. 633.

Le Sifilet, ou Manucode à six filets, Montbeill. Hist. Nat. Ois. iii. p. 156 (1774).—Forst., in Forrest, Voy. Molucq. et à la Nouv. Guin. p. 163 (1780).—Id. Ind. Zool. p. 38 (1781).

L'Oiseau de Paradis à gorge dorée, Sonnerat, Voy. Nouv. Guin. p. 158, pl. 97 (1776).

Paradisea sefilata, Pennant, Faunula Indica, in Forst. Zool. Ind. p. 40 (1781: ex D'Aubent.).

Golden-breasted Bird of Paradise, Lath. Gen. Syn. ii. p. 481 (1783).—Id. Gen. Hist. B. iii. p. 194, pl. xlvii. (1822).

Paradisea sexpennis, Bodd. Tabl. Pl. Enl. p. 38 (1783).—Gray, Gen. B. ii. p. 322 (1847).—Id. P. Z. S. 1858, p. 194.—Id. Cat. B. New Guinea, pp. 30, 50 (1859).—Id. P. Z. S. 1861, p. 436.—Schl. J. f. O. 1861, p. 385.—Id. Mus. Pays-Bas, Coraces, p. 92 (1867).—Id. Dierent. p. 173, cum fig. (c. 1870).—Id. Nederl. Tijdschr. Dierk. iv. pp. 42, 50 (1871).—Gray, Hand-l. B. i. p. 16, no. 6255 (1870).—Rosenb. Reist. naar Geelvinkb. p. 116 (1875).—Id. Malay Archip. p. 557 (1879).—Musschenbr. Dagboek, pp. 192, 225 (1883).—Rosenb. Mitth. orn. Ver. Wien, 1885, p. 40.

Paradisea aurea, Gm. Syst. Nat. i. p. 402 (1788).—Bechst. Kurze Uebers. p. 133 (1811).—Cuv. Règn. Anim. i. p. 404 (1817).—Dumont, Dict. Sc. Nat. xxxvii. p. 511 (1825).—Cuv. Règn. Anim. 2nd ed. i. p. 437 (1829).—Finsch, Neu-Guinea, p. 173 (1865).

Paradisea sexsetacea, Lath. Ind. Orn. i. p. 195 (1790).—Daud. Orn. ii. p. 276 (1800).—Shaw, Gen. Zool. vii. pt. 2, p. 490, pl. 65 (1809).—Ranz. Elem. Zool. iii. part 4, p. 77, tab. xlii. fig. 2 (1822).—Wagl. Syst. Av., Paradisea, sp. 6 (1827).—Less. Man. d'Orn. i. p. 364 (1828).—Wallace, P. Z. S. 1862, pp. 154, 157.

Le Sifilet, Vieill. Ois. Dor. ii. Ois. Parad. p. 18, pl. 6 (1802).—Le Vaill. Ois. Parad. i. pls. 12, 13, & 11 c, d, e, f (1806).—Less. Voy. Coq., Zool. i. pt. 2, p. 654 (1828).

Parotia sexsetacea, Vieill. N. Dict. d'Hist. Nat. xxxi. p. 160 (1819).—Id. Enc. Méth. ii. p. 205, pl. 144 fig. 5 (1822).—Id. Gal. Ois. i. p. 148, pl. 97 (1826).—Less. Traité d'Orn. p. 337 (1831).—Id. Illustr. Zool. pl. iv. (1831).—Id. Ois. Parad. , Syn. p. 10, Hist. Nat. p. 172, pls. x., xi. (e xd.), xi bis (e jun.), xii. (x) (1831).—Swains. Class. B. ii. p. 332 (1837).—Less. Compl. de Buff., Ois. p. 461 (1838).—Wall. P. Z. S. 1862, pp. 159, 160.—Rosenb. Nat. Tijdschr. Nederl. Ind. xxv. p. 247 (1863).—Id. J. f. O. 1864, p. 131.

Parotia aurea, Stephens, in Shaw's Gen. Zool. xiv. p. 75 (1826).—Gray, List Gen B. p. 39 (1840).—Id. ibid. 2nd ed. p. 52 (1841).—Bp. Consp. Av. i. p. 414 (1850).—Gray, List of Gen. & Subgen. p. 65 (1855).—Wall. Ibis, 1861, p. 287.

Parotia sexpennis, Scl. Journ. Linn. Soc. ii. p. 163 (1858).—Wall. Malay Archip. ii. pp. 608 (cum fig.), 619 (1869).—Elliot, Monogr. Parad., Introd. p. xix, pl. 10 (1873).—D'Alb. P. Z. S. 1873, p. 557.—Scl. t. c. p. 697.—Beccari, Ann. Mus. Gen. vii. p. 712 (1875).—Salvad. t. c. pp. 782, 699.—Gould, B. New Guinea, i. pl. 25 (1875).—Salvad. Ann. Mus. Gen. ix. p. 190 (1876), x. p. 155 (1877).—D'Alb. Nuova Guinea, pp. 70, 72, cum tab., & p. 582 (1880).—Salvad. Orn. Papuasia, etc. ii. p. 515 (1881).—Guillem. P. Z. S. 1885, p. 647.—D'Hamonv. Bull. Soc. Zool. France, xi. p. 519 (1886).—Salvad. Agg. Orn. Papuasia, etc. ii. p. 149 (1890).—Sharpe, Bull. Brit. Orn. Club, iv. p. xiv (1894).

Lophorhina sexpennis, Sundev. Meth. Nat. Av. Disp. Tent. p. 45 (1872).

Parotia sefilata, Sharpe, Cat. Birds Brit. Mus. iii. p. 177 (1877).—Eudes-Deslongch. Ann. Mus. d'Hist. Nat. Caen, p. 35 (1880).

This is a species of Bird of Paradise which has been known to science for the last hundred and thirty years, having been first figured by D'Aubenton in the 'Planches Enluminées,' under the name of "Le Sifilet." On this French name Pennant founded his *Paradisea sefilata*, which is really the oldest name of the species; but Count Salvadori, who has given the most complete synonymy of this species, says that he lacks the courage to adopt such a word as *sefilata*, a barbarous Latin translation of the French 'Sifilet.' For the same reason I follow the Count in this respect and adopt Boddaert's name, which is next in order of date.

An excellent account of the changes of plumage of this Bird of Paradise is given by Count Salvadori in his 'Ornitologia della Papuasia.' A large series of specimens were brought from North-western New Guinea by Signor D'Albertis and Dr. Beccari, as well as by Mr. Bruijn's native hunters, and from this series Count Salvadori gives the following account of the sequence of plumages undergone by the males. These at first resemble the adult female, from which it is impossible to distinguish them, but the frontal plumes gradually increase in density and then assume the silvery white forehead which is so conspicuous in the adult. Then are assumed gradually the metallic colours of the occipital diadem, and at the same time some metallic feathers also appear on the throat and fore-neck, while the brown plumage becomes black and of a velvety texture. Successively the feathers of the wings and tail begin to be black and present a velvet aspect, while a collar of velvety black plumes appears on the neck, the feathers of which are lengthened. The metallic shield on the throat gradually extends, as does also the black of the chin and the sides of the breast; the breast, abdomen, and under tail-coverts and the long flank-feathers also blacken, and the three long plumes on each side of the crown make their appearance, being black from the very commencement. The last portions of the body to show remains of young plumage are the abdomen and the rump. The series of specimens in the British Museum has not been extensive enough to allow me to follow all the changes of plumage described by Count Salvadori, but the allied species, P. lawesi, appears to go through a very similar ascension; but the order of change appears to me to vary with individuals. Thus some specimens commence to develop the ornamental plumes on the head before any sign of the gular shield is apparent, and they are occasionally fully developed when only a slight blackening of the throat has commenced. The rackets at the end of the ornamental plumes are at first longitudinal in shape, but quickly assume their ovate form.

The home of this species is the Arfak Mountains, in North-western New Guinea. In the Charles Lewis Mountains it is represented by P. carola, and in the Owen Stanley Mountains by P. lawesi.

Dr. Beccari has given the following note on the species:—" Of Parotia sexpennis I got one adult male alive, but it only lived three days. Its eye, with the iris azure surrounded by a yellow ring, is extremely beautiful. The six feathers which ornament the head are not raised up vertically, but moved backwards and forwards in a horizontal and oblique direction, and are moved forward parallel to the sides of the beak. It is the commonest Paradise-bird at Mount Arfak, but, as usual, the adult males are much scarcer than the females and young males."

The following remarks are from the pen of Signor D'Albertis:—" Although this species has been known for many years, it is not yet accurately understood, having only been described from birds in a mutilated condition. My observations have been made in the natural haunts of these elegant birds, from numerous specimens both living and dead. These birds are found in the north of New Guinea. I met with them about thirty miles from the coast, at an elevation of 3000 feet above the level of the sea, near Mount Arfak. I have never found the adult male in company with females or young birds, but always in the thickest parts of the forest; the females and young birds are generally found in a much lower zone. This Paradise-bird is very noisy, uttering a note like ' gnand-gnand.' It feeds upon various kinds of fruits, more especially a species of fig which is very plentiful in the mountain-ranges; at other times I have observed it feeding on a small kind of nutmeg. To clean its rich plumage this bird is accustomed, when the ground is dry, to scrape (similarly to a gallinaceous bird) around places clear of all grass and leaves, and to roll over and over again in the dust produced by the clearing, at the same time crying out, extending and contracting its plumage, elevating the brilliant silvery crest on the upper part of its head, and also the six remarkable plumes from which it derives the specific name of sexpennis. On seeing its eccentric movements at this time, and hearing its cries, one would consider it to be engaged in a fight with some imaginary enemy. This bird is named ' Camu-o ' by the natives. I have also a skeleton of a young male of this species, which, although not in a perfect state, may no doubt be interesting as showing the form of the cranium, on which there is an admirable muscular structure which enables the bird to elevate the feathers of the head. The feathers at the nape of the neck exhibit, when the rays of light strike upon them, a rich and brilliant metallic hue. The eyes are of a light blue, with a circle of pale yellowish-green colour."

Adult male. General colour above rich purplish velvety black, including the wings and tail; plumes of head and neck also close-set and velvety, purplish black like the back; above the ear-coverts a tuft of elongated silky hair-like plumes, from among the anterior ones of which spring three thread-like shafts on either side of the head, each ornamented with an ovate racket of velvety black; frontal plumes purplish, the feathers tipped with shining white, which forms a band across the forehead, these stiff plumes being capable of depression forwards nearly to the tip of the bill; across the upper crown a band of brilliant metallic plumes, the centre ones bright green with a double sub-terminal bar of purple and blue, the outermost ones somewhat more shining with metallic purple and blue; throat velvety purplish black, the lower throat and fore-neck forming a brilliant metallic shield, composed of golden-copper feathers, shining with steel-green, blue, and purple, and relieved by a median spot of velvety black ; rest of under surface of body glossy black ; on each side of the breast the plumes elongated into two lax velvety shields : bill and legs black. " iris light blue, with a circle of pale yellowish-green colour " (*D'Albertis*). Total length 12·5 inches, culmen 1·3, wing 6·1, tail 5·25, tarsus 2·1.

Adult female. Differed from the male, being dark chestnut-brown above ; quills and tail blackish, externally reddish brown ; crown of head, nasal plumes, feathers round the eye, and ear-coverts black ; cheeks and throat greyish white, finely barred with black and separated from each other by a broad malar band of black ; sides of neck black, encroaching on the sides of the lower throat ; remainder of under surface of body, from the lower throat downwards, isabelline, regularly barred with black. Total length 11·5 inches, wing 5·9.

The specimens figured in the Plate are in the British Museum, and the descriptions are taken from birds in the same collection.

PAROTIA LAWESI, *Ramsay.*

Southern Six-plumed Bird of Paradise.

Parotia lawesi, Ramsay, Proc. Linn. Soc. N. S. Wales, x. p. 243 (1885).—Id. Nature, vol. xxxii. p. 298 (1885).—Finsch & Meyer, Zeitschr. ges. Orn. ii. p. 375, Taf. xvi. (1885).—Id. Ibis, 1886, p. 243.—D'Hamonv. Bull. Soc. Zool. France, xi. pp. 507, 510 (1886).—Sharpe, Nature, xxxiv. p. 340 (1886).—Id. in Gould's B. New Guinea, i. pl. 26 (1887).—Salvad. Agg. Orn. Papuasia e delle Molucche, ii. p. 149 (1890).—Goodwin, Ibis, 1890, p. 153.

Parotia sexpennis (nec Bodd.), De Vis, Ibis, 1891, p. 93.

Tais species has been named by Dr. E. P. Ramsay in honour of the Rev. W. G. Lawes, a well-known missionary in South-eastern New Guinea, who has taken great interest in the natural history of that part of the world, and has made some collection of the animals and plants of S.E. Papua. It represents in the Astrolabe Range the Six-plumed Bird of Paradise of North-western New Guinea, but differs from that species in the colouring of both sexes.

Dr. Ramsay was acquainted only with the male, but both sexes were discovered by the late Karl Hunstein in the Horseshoe Range of the Owen-Stanley Mountains, and they were figured in the celebrated paper of Dr. Otto Finsch and Dr. A. B. Meyer, published in the second volume of Madarász's "Zeitschrift für die gesammte Ornithologie." Afterwards Mr. H. O. Forbes met with the species in the Sogeri district of the Astrolabe Range, and during Sir William Macgregor's expedition into the interior of the Owen-Stanley Mountains the Six-plumed Bird of Paradise was found on Mount Belford, at an altitude of from 3000 to 7000 feet. Mr. De Vis, who has described the collections made during this expedition, does not admit the distinctness of *P. lawesi* from *P. sexpennis*, and records the southern species under the latter name.

Mr. A. P. Goodwin, who was one of the naturalists who accompanied Sir William Macgregor, states that he first heard the call of *Parotia lawesi* on Mount Belford at the altitude of 4000 feet. He adds:—" I did not succeed in obtaining a specimen until we had crossed the Joseph River and had ascended Mount Musgrave to the same altitude. Here I secured an example of this beautiful bird. Our camp was near one of their play-grounds, so I had a good opportunity of watching the bird's movements. It has a strong resemblance to the Silky Bower-bird (*Ptilonorhynchus holosericeus*) of New South Wales both in form and habits. It has a similar bill, beautiful blue eyes, and strong legs, and, like the Bower-bird, is very cautious, restless, and swift. It has also a similar flight. Although *P. lawesi* does not build a bower, still it has its play-ground, where a number of these birds (from six to eight) may be found playing together. The play-ground may be easily known by the colour of the soil and by the clearance of the surrounding underbrush."

The British Museum is indebted to the Hon. Hugh Romilly for a series of skins of this Bird of Paradise, and examples of both adult and young birds are now represented in the collection.

Adult male. Similar to *P. sexpennis*, but a little smaller, and differing in the colour of the metallic pectoral shield, which is of a fiery golden lustre, with less green than is shown by *P. sexpennis*. The metallic band on the nape is purplish, with steel-blue reflections, with only a faint lustre of green near the base, whereas *P. sexpennis* has the centre of this nape-band green. The silvery white patch on the forehead comes to an obtuse point at the base of the bill, forming a triangle on the forehead, whereas in *P. sexpennis* the forehead is velvety black, succeeded by a fan-shaped band of silvery white. Total length 13 inches, culmen 1·15, wing 6·15, tail 5·0, tarsus 2·15.

Adult female. Similar to the female of *P. sexpennis*, but rather smaller, more chestnut above, and easily distinguished by the colour of the under surface, which is rufous barred with black, whereas in *P. sexpennis* the under surface is silvery grey. Total length 9·5 inches, culmen 1·1, wing 6·1, tail 4·1, tarsus 2·0.

The young male at first resembles the adult female.

A pair of adult birds are figured in the Plate, from specimens procured by Hunstein on the Horseshoe Range of the Astrolabe Mountains.

PAROTIA CAROLÆ, *Meyer.*

J. G. Keulemans & Hart del. et lith.   Mintern Bros. imp.

PAROTIA CAROLÆ, Meyer.

Queen of Saxony's Bird of Paradise.

Parotia carolæ, Meyer, Bull. Brit. Orn. Club, iv. p. vi (Nov. 1894).—Sharpe, Bull. Brit. Orn. Club, iv. p. xlvi (Dec. 1894).—Meyer, Abh. k. zool. Mus. Dresden. 1894-95, no. 8, p. 8, Taf. ii. (1895).—Büttikofer, Notes Leyden Mus. xvii. p. 37 (1895).—Rothschild, Bull. Brit. Orn. Club, iv. p. xxii (1895).

This beautiful species of Six-plumed Bird of Paradise was described by Dr. A. B. Meyer from the mountains of Amberno, in North-western New Guinea, to the eastward of Geelvink Bay. Since the arrival of the first specimen in Europe, a considerable number of skins have been received, and the species is now represented in most of the public and private museums of Europe.

There is no need to enlarge on the differences between *Parotia carolæ* and the other two species of the genus *Parotia*. Besides the white plumes on the flanks and the difference in the crest, the six ornamental plumes on the sides of the crown carry very small rackets.

The form of the crest in this species is also peculiar, and the skins first sent to Europe had the crest-feathers so pressed together that the white-tipped plumes hid the median tuft of whitish frontal feathers. This was pointed out by Dr. Büttikofer in 1895, and an illustration of the true form of the crest was given by him (*l. c.*).

P. carolæ has the first two primaries sinuated at the end, very much as in *P. lawesi*, but with a loop in the inner web at the commencement of the sinuation. The tail is more square than in *P. lawesi*, and not at all like the graduated tail of *P. sexpennis*.

Adult male. Velvety black, with purplish-violet reflections; wings velvety black, the quills dull black, the first two primaries having a terminal attenuation; tail velvety black; sides of the crown velvety black, the feathers tipped with silvery white, which forms a lateral crest; this crest when closed covering the centre of the crown, which commences with a crinkled patch of silvery white feathers on the forehead, followed by a patch of glittering golden plumes; on the nape a triangular shield of metallic feathers, first green, then steel-blue, and lastly purplish violet; on each side of the occiput six thread-like plumes ending in a very small racket; round the eye golden chestnut; sides of the neck velvety black, the feathers elongated and forming a frill; cheeks golden sienna; throat dull ashy whitish, mottled with dusky bases to the feathers, the chin blackish; on the fore-neck a triangular shield of metallic lilac with reflections of golden bronze, the feathers centred with spots of velvety black; remainder of under surface of body black, the sides of the body ornamented with long white plumes, or half chestnut and half white, while some are chestnut with black centres; under wing-coverts black, with white and chestnut streaks. Total length 11 inches, culmen 0·75, wing 6·3, tail 2·6, tarsus 2.

The female has not as yet been described, but it is evidently a reddish-brown bird, barred with black, to judge from the plumage of the young male, which has the breast reddish brown, crossed with black bars.

The Plate represents an adult male and a young male, drawn from specimens in the British Museum.

PAROTIA BERLEPSCHI, *Kleinschm.*

Count von Berlepsch's Bird of Paradise.

Parotia berlepschi, Kleinschmidt, Orn. Monatsb. v. p. 46 (1897).—Id. J. f. O. 1897, p. 174.

This is an ally of *Parotia carolæ*, from which it only differs in some few characters. On the back there is a wash of golden brown; the commissure of the bill is more curved, and the bill is higher than in *P. carolæ*. The white frontal patch is also smaller than in the latter species, and the lateral feathers of the coronal crest lack the white tips; but the most striking peculiarity of *P. berlepschi* seems to me to be the colour of the cheeks and throat, which, instead of being of light ochreous or golden colour, are blackish with a faint reddish-brown gloss under certain lights.

The exact part of New Guinea from which *P. berlepschi* comes is at present unknown, and only a few native-prepared skins have so far come to Europe and are now in the collections of Count von Berlepsch and the Hon. Walter Rothschild.

The similarity of *P. berlepschi* to *P. carolæ* has rendered a separate figure of the species unnecessary.

PAROTIA HELENÆ, *De Vis.*

Mount Scratchley Bird of Paradise.

Parotia helenæ, De Vis, Ibis, 1897, p. 380.

I HAVE not seen a specimen of this recently discovered Bird of Paradise, which can be best described by quoting the original account given by Mr. De Vis :—

"This species bears a very close general resemblance to *Parotia lawesi* of Ramsay, and might be described in the same terms, were it not differentiated from it by the form and colouring of the crest and the non-elongation of the superciliary plumes ; the supranasal part of the crest is erect and very low anteriorly and ascends gently to the forehead ; the frontal part is suddenly elongated and forms a compressed rounded lobe ; the short anterior portion is bright bronze-brown ; the elevated posterior part is dark coffee-brown, with a paler bronze-brown reflection, and the adjacent parts of the head are similar in colour and lustre : " iris in both sexes yellow, pupil light blue ; beak black ; feet in male corneous or light brown, in female iron-grey Contents of stomach, fruits. Native name 'Kamaro.'"

"The females of the two species can be distinguished only by the colour of the thighs, which in *P. lawesi* are rufous, in *P. helenæ* fuscous."

Four males and a female were obtained at Neneba, on Mount Scratchley, in South-eastern New Guinea, at a height of 4000 feet, in November 1896.

LOPHORHINA SUPERBA (Penn.).

Superb Bird of Paradise.

Le Manucode noir de la Nouvelle Guinée, dit le Superbe, Montb. Hist. Nat. Ois. iii. p. 187 (1774).—Forster, Zool. Ind. p. 38 (1781).

Oiseau de Paradis de la Nouvelle Guinée, dit le Superbe, D'Aubent. Pl. Enl. iii. pl. 634.

L'Oiseau de Paradis à gorge violette, surnommé le Superbe, Sonn. Voy. Nouv. Guin. p. 157, pl. 96 (1776).

Paradisea superba, Penn. Faunula Ind. in Forst. Ind. Zool. p. 40 (1781).—Scop. Del. Flor. et Faun. Insubr. ii. p. 88 (1786).—Gm. Syst. Nat. i. p. 401 (1788).—Lath. Ind. Orn. i. p. 196 (1790).—Shaw, Gen. Zool. vii. pt. 2, p. 494, pls. 63, 64, 65 (1809).—Id. & Nodder, Nat. Misc. xxvi. pl. 1021.—Cuvier, Règne Anim. i. p. 404 (1817).—Raoz. Elem. Zool. iii. part 4, p. 72, tav. xiv. fig. 1 (1822).—Dumont, Dict. Sci. Nat. xxxvii. p. 510 (1825).—Wagler, Syst. Av. Paradisea, p. 102, sp. 5 (1827).—Less. Man. d'Orn. i. p. 394 (1828).—Drap. Dict. Class. xiii. p. 47 (1829).—Cuvier, Règne Anim. i. p. 488 (1829).—Wallace, Ibis, 1859, p. 111.—Id. Proc. Zool. Soc. 1862, p. 154.—Finsch, Neu-Guinea, p. 173 (1865).—Rosenb. Reist. naar Geelvinkb. p. 110 (1875).

Superb Paradise Bird, Lath. Syn. Av. ii. p. 479 (1782).—Id. Gen. Hist. Birds, iii. p. 191 (1822).

Paradisea atra, Bodd. Tabl. Pl. Enl. p. 38 (1783).—Gray, Gen. B. ii. p. 322 (1847).—Id. Proc. Zool. Soc. 1858 p. 184.—Id. Cat. B. New Guinea, pp. 36, 56 (1859).—Id. Proc. Zool. Soc. 1861, p. 436.—Schl. J. f. O. 1861, p. 385.—Id. Dieront. p. 173 (c. 1870).—Rosenb. Malay. Arch. p. 558 (1879).—Musschenbr. Dagbeek, pp. 193, 226 (1883).—Rosenb. Mitth. orn. Ver. Wien, 1884, p. 40.

Paradisea fuscata, Lath. Ind. Orn. i. p. 196 (1790).

Le Superbe, Vieill. Ois. Dor. ii. Ois. Parad. pl. vii. (1802).—Le Vaill. Ois. Parad. i. pls. 14, 15 (1806).—Less. Voy. Coquille, Zool. i. pt. 2, p. 634 (1828).

Lophorina superba, Vieill. Analyse, p. 35 (1816).—Id. Nouv. Dict. d'Hist. Nat. xviii. p. 165 (1817).—Id. Enc. Méth. p. 910, pl. 163, fig. 4 (1823).—Id. Gal. Ois. i. p. 149, pl. xcviii. (1825).—Steph. Gen. Zool. xiv. p. 56 (1826).—Less. Traité, p. 337 (1831).—Id. Ois. Parad. Syn. p. 12.—Id. Hist. Nat. p. 179, pls. xiii., xiv. (1835).—Less. Compl. Buff. Ois. p. 463 (1838).—Wall. Ibis, 1861, p. 287.—Id. Proc. Zool. Soc. 1862, pp. 160, 160.—Rosenb. Nat. Tijdschr. Nederl. Ind. xxv. p. 131 (1863).—Id. J. f. O. 1864, p. 347.—Salvad. Ann. Mus. Genov. ix. p. 190 (1876), x. p. 155 (1877).—Sharpe, Cat. Birds Brit. Mus. iii. p. 179 (1877).—Gould, Birds of New Guinea, i. pl. 18 (1878).—Vaillant, Ann. Sci. Nat. (6) viii. art. 13, pls. x., xi. (1879).—D'Albert. Nuova Guinea, pp. 57, 71, 509, cum tab. (1880).—Salvad. Orn. Papuasia, etc. ii. p. 574 (1881).—Finsch-Deningeth, Ann. Mus. Caen. i. p. 37 (1889).—Guillem. Proc. Zool. Soc. 1885, p. 617.—Salvad. Agg. Orn. Papuasia, etc. ii. p. 150 (1890).

Lophorhina atra, Sclater, Journ. Linn. Soc. ii. p. 163 (1858).—Wallace, Malay Arch. ii. p. 496, cum fig. (1869).—Gray, Hand-l. B. ii. p. 18, no. 6254 (1870).—Sundev. Av. Tent. p. 45 (1872).—Elliot, Monogr. Parad. pl. 11 (1873).—D'Albert. Proc. Zool. Soc. 1873, p. 558.—Sclater, Proc. Zool. Soc. 1873, p. 697.—Meyer, Mitth. k. Zool. Mus. Dresden, i. pp. 7, 8 (1875).—Beccari, Ann. Mus. Genov. vii. p. 713 (1875).—Salvad. tom. cit. p. 783 (1875).—D'Albert. tom. cit. p. 798.—Id. Nuova Guinea, pp. 67, 311 (1880).—Cory, Beautiful and Curious Birds, part vi. (1883).—D'Hamonv. Bull. Soc. Zool. France, 1886, p. 510.

Epimachus ater, Schlegel, Mus. Pays-Bas, Coraces, p. 95, note (1876).—Id. Nederl. Tijdschr. Dierk. iv. p. 17 (1871).

This beautiful Bird of Paradise is confined to New Guinea, and has a somewhat restricted range, as, so far as we know, it is only found in the north-western portion of the island. In fact it is as yet known only from the Arfak Mountains, being replaced by *Lophorhina minor* in the mountains of South-eastern New Guinea. In the former locality the species has been found by D'Albertis and Beccari, and many specimens procured by Mr. Bruijn's hunters have found their way to Europe.

Signor D'Albertis says :—" The natives give this bird the name of ' Nieddo,' derived from the sound of its notes ! "

Dr. Beccari does not give much information about the present species in his " Ornithological Letter " from North-western New Guinea. He merely observes :—" *Lophorhina atra* is rather rarer than *Paradisea*; but I must tell you that the abundance of fruit-eating birds in a given locality depends principally on the season at which certain kinds of fruit are ripe; therefore a species may be common in a place one month, and become rare or completely disappear in the next, when the season of the fruit on which it lives has passed."

The following description is from the third volume of the 'Catalogue of Birds' :—

Adult male. General colour above velvety black, somewhat glossed with bronzy purple ; mantle produced into an elevated shield, composed of velvety black plumes, glossed under certain lights with bronze ; wing-coverts velvety black, rather more distinctly glossed with purple than the back ; quills and tail-feathers deep black, glossed with steel-blue ; lores and nasal plumes forming an elevated crest of purplish-black feathers ; crown of head, nape, and hind neck spangled with metallic steel-coloured feathers, each of which has a sub-terminal bar of purple ; sides of face, sides of neck, and entire throat deep coppery bronze ; on the fore neck and breast a pectoral shield of bright metallic green plumes, most of which have a narrow edging of copper ; remainder of under surface purplish black. Total length 9 inches, culmen 1·15, wing 4·55, tail 3·6.

Adult female. Above deep chocolate-brown, the feathers of the top and sides of the head blackish brown ; over the eye a few white-spotted plumes ; wing-coverts and quills blackish brown, externally reddish ; tail brown, externally dull rufous brown ; throat white, all the feathers being black tipped with white ; rest of the under surface buffy white, inclining to rufous on the flanks and under tail-coverts, the whole under surface barred across with dull brown ; under wing-coverts rufous, barred across with brown. Total length 8·9 inches, culmen 1·05, wing 3·1, tail 4, tarsus 1·3.

The principal figure is of the natural size, with a reduced male and female in the distance.

LOPHORHINA MINOR, Ramsay.

Lesser Superb Bird of Paradise.

Lophorhina superba minor, Ramsay, Proc. Linn. Soc. N. S. Wales, x. p. 242 (1885).

Lophorhina minor, Finsch & Meyer, Zeitsche. ges. Orn. ii. p. 376, pl. xvii. (1885).—Id. Ibis, 1886, p. 244.—Meyer, op. cit. iii. p. 191, cum fig. (1885).—D'Hamonv. Bull. Soc. Zool. France, 1886, pp. 508, 510.—Sharpe, in Gould's Birds of New Guinea, i. pl. 19 (1889).—Salvad. Agg. Orn. Papuasia, ii. p. 150 (1890).—Goodwin, Ibis, 1890, p. 152.—Salvad. Agg. Orn. Papuasia, iii. p. 240 (1891) = Sharpe, Bull. Brit. Orn. Club, iv. p. xvi (1894).

Lophorhina superba (nec Penn.), De Vis, Colonial Papers, no. 193, p. 113 (1889).—Id. Annual Report British New Guinea, p. 60 (1889).—Id. Ibis, 1891, p. 36.

This species is a smaller representative of the Superb Bird of Paradise, *Lophorhina superba*, of North-eastern New Guinea, and was discovered by the late Mr. Carl Hunstein in the Astrolabe Range in South-eastern New Guinea.

The form of its neck-shield is, however, quite different, as was discovered by Dr. A. B. Meyer, when he had a specimen mounted for the Gallery of the Dresden Museum. Numerous specimens have been sent to the British Museum by the late Hon. Hugh Romilly, and by Dr. H. O. Forbes, who procured the species in the Sogeri district. Mr. Goodwin, who accompanied the expedition of Sir William Macgregor to the Owen Stanley Mountains, writes to me:—"At an altitude of 5000 feet we came across this Superb Bird of Paradise, and as it fluttered about on the highest perch it could find it looked no bigger than a butterfly. Needless to say, but few specimens were obtained. Its call resembles that of *Parotia lawesi*, but is not so loud."

Sir William Macgregor records the species on Mount Owen Stanley at 4350 feet, and again from Goodwin Spur, at from 5000 to 7000 feet. It is not noticed in Dr. Meyer's account of the collections from Kaiser Wilhelm's Land, so that it seems to be entirely a species of the Owen Stanley Mountains, so far as is yet known.

On re-examining the specimen of *L. superba* figured by Gould, and comparing it with the specimens of *L. minor* in the British Museum, from South-eastern New Guinea, not only is the shape of the cervical shield found to be different, as pointed out by Dr. Meyer, but the plumes overhanging the base of the bill may also prove to be differently disposed.

In *L. superba* the angle of the chin is covered up by velvety plumes, and above the nostrils the feathers widen out into a kind of small fan. The arrangement of the feathers of this part of the head is, however, one which is directly affected by the process of preparing the skin, and I suspect that there is really no difference in life between the two species in respect to the arrangement of the plumes on the nose and on the chin. In *L. minor* the shield is not nearly so dense, the fork in the middle is much more marked, and the lateral feathers are rounded on the ends and not so pointed as in *L. superba*.

Adult male. General colour velvety black, with reflexions of coppery bronze on the mantle and cervical shield, the feathers of these ornamental plumes being edged with oily green at the ends; back and rump duller black, the upper tail-coverts velvety black, glossed with purple, the centre feathers with violet-blue; crown of head metallic steel-green, with a few metallic purple feathers on the nape; sides of face and throat velvety black with an oily green shade; across the fore-neck a brilliant elongated shield of metallic bluish green; rest of under surface of body black. Total length 8·3 inches, culmen 1, wing 5·3, tail 3·2, tarsus 1·3.

The adult female is very similar to the male, is much lighter brown and not so chestnut as the female of *L. superba*, and appears to differ also in having a line of white feathers dotted with black from the hinder part of the eye above the ear-coverts; the outer aspect of the quills is paler rufous than the dark chestnut of the wing in *L. superba*, and the tail is olive-brown. The under surface of the body is paler buff, and the cross-bars are paler.

The descriptions have been taken from a pair of birds in the British Museum, and the figures on the Plate were drawn from the same birds.

LOBOPARADISEA SERICEA, *Rothschild.*

J.G. Keulemans & J.W. Hart del. et lith.

LOBOPARADISEA SERICEA, *Rothschild.*

Shield-billed Bower-bird.

Loboparadisea sericea, Rothschild, Bull. Brit. Orn. Club, vi. p. xv (1896).—Id. Novit. Zool. iv. p. 169, pl. ii. fig. 2 (1897).

THE only specimen at present known of this curious bird is in the collection of the Hon. Walter Rothschild, at Tring. It has been described by him as a Bird of Paradise, but it is apparently a Bower-bird, though this is a question difficult to settle at the present moment. From a comparison of its characters it would appear to be related to *Loria* and *Cnemophilus*, as it has the nasal aperture covered by a wattle, in place of the feathers which hide the nasal opening in the two above-named genera. In the true Bower-birds the nasal aperture is exposed.

Mr. Rothschild writes:—"The colour of the wattles is guessed from what they look like in the dried skin, which is said to have been bought from natives at Kocroedoe, on the northern coast of Dutch New Guinea. This place, Kocroedoe, is not to be mistaken for Korrido in Geelvink Bay."

The following is a description of the type specimen in Mr. Rothschild's collection:—

General colour above chestnut-brown, with a slight golden shade on the hind-neck and mantle; wings rather more chestnut than the back; quills chestnut-brown, with dusky tips to the inner webs, decreasing in extent on the secondaries, which are almost entirely reddish brown; lower back and rump sulphur-yellow; upper tail-coverts and tail chestnut-brown; crown of head and nape dusky brown, contrasting slightly with the back; the sides of the face darker than the head; cheeks and under surface of body sulphur-yellow; the under tail-coverts tipped with chestnut; thighs reddish brown; axillaries sulphur-yellow, slightly washed with chestnut; under wing-coverts and quill-lining chestnut: "bill with two large wattles reaching halfway down from the base, dull blue with yellow tips" (*W. Rothschild*). Total length 6·5 inches, culmen 0·75, wing 3·5, tail 2·1, tarsus 1·2.

The figure represents the type specimen of the size of life, drawn from a painting by Mr. Keulemans. I have to acknowledge Mr. Rothschild's kindness in permitting me to describe and figure the specimen.

LORIA MARIE (De Vis).

LORIA MARIÆ (*De Vis*).

Lady Macgregor's Bower-bird.

Cnemophilus mariæ, De Vis, Annual Report British New Guinea, 1893–94, p. 104 (1894).—Sharpe, Bull. B. O. Club, iv. p. xiv (1894).

Loria mariæ, Sclater, Ibis, 1895, p. 343, pl. viii.—Rothschild, Nov. Zool. iii. p. 14 (1896).

Loria loriæ, Rothschild, Nov. Zool. iii. p. 252 (1896).

THIS interesting species was discovered on Mount Maneao, in South-eastern New Guinea, during Sir William Macgregor's exploration of this mountain in 1894, by Captain Armit and Mr. Guise, who accompanied the expedition. The species was named after Lady Macgregor by Mr. C. W. De Vis, and he afterwards very kindly sent the type specimens over to England to Dr. Sclater, who figured them in the ' Ibis ' for 1895. These specimens have also formed the subject of my Plate in the present work, but I have endeavoured to give a more satisfactory rendering of the colours of this species, as the plate in the ' Ibis ' is, I regret to say, not correct.

In May 1894, Count Salvadori described as a new genus and species of the *Paradiseidæ* a bird from the Moroka district in the Owen Stanley range. Only a female was sent by Dr. Loria; but Count Salvadori recognized that it represented a new form, and he named it after its discoverer, *Loria loriæ*. When Mr. De Vis sent over the types of his *Cnemophilus mariæ* to Dr. Sclater for examination, Count Salvadori was so good as to allow the type of his *Loria loriæ* to be sent for comparison, and I had the pleasure of comparing these two species together. I fully agreed with Dr. Sclater that *C. mariæ* was a *Loria*, and, like him, I could not advise that the two species *L. loriæ* and *L. mariæ* were identical, because neither of the females sent by Mr. De Vis showed the naked line of yellow skin from the gape to below the ear-coverts which the type of *Loria loriæ* so strongly exhibited, and on which character Count Salvadori laid emphasis in describing the genus. I have therefore kept the name of *Loria mariæ* for the Mount Maneao bird.

More recently, however, the Hon. Walter Rothschild has obtained four specimens of the genus *Loria*. A female was obtained by his collector in the Eafa district of the Owen Stanley range, between Mounts Alexander and Bellamy. Mr. Rothschild also has a beautiful male specimen from the Sakestasuma range in the Kaiari district, between Mounts Alexander and Nisbet, as well as a young male from Mount Victoria. Although the female bird does not show the bare oral streak as in the type of *L. loriæ*, I perceive indications of it in Mr. Rothschild's specimen, and I expect that his later conclusion will prove to be correct, that *L. mariæ* is identical with *L. loriæ*, and that Count Salvadori's name will have to stand for the species. Mr. Rothschild has also a trade-skin, said to have come from the Arfak district of North-western New Guinea, which is certainly identical with the Owen Stanley specimens.

Adult male. General colour above velvety black with a purplish gloss; wing-coverts also velvety black, the quills likewise velvety in texture, but, when held away from the light, the inner secondaries appear of a beautiful metallic steel-blue, glossed with purple; tail-feathers velvety black, with a metallic purple shade under certain lights; head exactly like the back, but the nasal plumes, lores, and a patch of feathers extending above the fore part of the eye metallic, changing under the light to greenish grey, emerald-green, or steel-green, sometimes showing a slight purplish tinge; sides of face and under surface of body velvety black, with a purplish gloss like the upper surface; bill black; feet dark green; iris brown, eyeball blue. Total length 8·5 inches, culmen 0·9, wing 4, tail 2·85, tarsus 1·5.

Adult female. Different from the male. General colour above olive-greenish, the wing-coverts like the back, but the greater series with a slight tinge of orange-brown; quills dusky brown, with a strong tinge of orange-brown externally, more bronzy on the secondaries; tail-feathers dusky brown, externally bronzy brown and margined with olive-greenish like the back; head like the back, slightly brighter and clearer olive on the lores and above the eye; sides of face and throat and chest olive-greenish, like the sides of the body; the breast, abdomen, and under tail-coverts lighter and more olive-yellow; under wing-coverts light tawny; quills ashy below, with the inner webs light tawny; bill black; feet green; iris greyish black. Total length 8·2 inches, culmen 0·9, wing 3·8, tail 2·65, tarsus 1·4.

A young male, passing to the adult plumage, is in Mr. Rothschild's collection. It is nearly velvety black like the adult, but still retains a considerable amount of the first plumage, which must have been like that of the adult female. The metallic inner secondaries, when first developed in the male, appear to be of a deeper blue than in the adults, while the metallic patch above the eye seems to be more purple than in the full-plumaged male.

The Plate represents an adult male and female of the size of life, and the figures are drawn from the type specimens kindly lent to me by Mr. De Vis. The descriptions are taken from the specimens in the Tring Museum, and for the opportunity of describing them I have to present my best acknowledgments to the Hon. Walter Rothschild.

PTILONORHYNCHUS VIOLACEUS (*Vieill.*)

Satin Bower-bird.

Pyrrhocorax violaceus, Vieill. Nouv. Dict. d'Hist. Nat. vi. p. 569 (1816).
Ptilonorhynchus holosericeus, Kuhl, Beitr. Zool. p. 159 (1820).—Wagler, Syst. Av., *Ptilonorhynchus*, sp. 1, p. 369 (1827).—Gould, B. Australia, iv. pl. 10 (1841).—Id. Handb. B. Austr. i. p. 442 (1865).—Ramsay, Ibis, 1866, p. 330.—Sclater & Wolf, Zoological Sketches, 2nd series, pl. xxviii. (1867).—Gray, Handlist Birds, i. p. 294, no. 4335 (1869).—Ramsay, Proc. Zool. Soc. 1875, p. 112.
Satin Grakle, Lath. Gen. Hist. iii. p. 171 (1822).
Kitta holosericea, Temm. Pl. Col. iv pls. 395, 422 (1826).—Lesson, Traité d'Orn. p. 350, pl. 46. fig. 1 (1831).
Ptilonorhynchus macleayi, Vig. & Horsf. Trans. Linn. Soc. xv. p. 263 (1827 : ex Lath. MSS.).
Ptilonorhynchus squamulosus, Wagl. Syst. Av., *Ptilonorhynchus*, sp. 3, p. 369 (1827 : ex Illiger, MSS.).
Ptilorhynchus holosericeus, Swains. Classif. B. ii. p. 271 (1837).—Gray, Gen. Birds, ii. p. 325 (1846).—Bp. Consp. Av. i. p. 370 (1850).—Cab. Mus. Hein. Th. i. p. 213 (1850).—Schl. Mus. Pays-Bas, Coraces, p. 117 (1867).—Ramsay, Proc. Linn. Soc. N. S. W. ii. p. 187 (1878).
Ptilorhynchus violaceus, Elliot, Monogr. Parad. pl. xxvii. (1870).
Ptilonorhynchus violaceus, Sharpe, Cat. Birds Brit. Mus. vi. p. 393 (1881).—North, Proc. Linn. Soc. N. S. W. (2) i. pp. 1155, 1171 (1887).—Ramsay, Tab. List Austr. B. p. 31 (1888).—North, Proc. Linn. Soc. N. S. W. (2) iii. p. 1776 (1889).—Id. Descr. Cat. Nests & Eggs Austr. B. p. 175, pl. xi. fig. 6 (1889).—Cox & Hamilton, Proc. Linn. Soc. N. S. W. (2) iv. p. 411 (1890).—Sharpe, Bull. Brit. Orn. Club, iv. p. xiv (1894).

THE present species is the best known of all the Bower-birds, not only from its common occurrence in Australia, but also on account of its having been frequently seen in our Zoological Gardens, where it constructs in captivity those wonderful bowers with which its name is associated.

The distribution of the Satin Bower-bird in Australia is extensive, reaching as it does from Rockingham Bay and the Port Denison district to New South Wales, Victoria, and South Australia.

The late Mr. John Gould gave an excellent account of the Satin Bower-bird in his "Birds of Australia":—

"It is a stationary species, but appears to roam from one part of a district to another, either for the purpose of varying the nature, or of obtaining a more abundant supply of food. Judging from the contents of the stomachs of the many specimens I dissected, it would seem that it is altogether frugivorous, or if not exclusively so, that insects form but a small portion of its diet. Independently of numerous berry-bearing plants and shrubs, the brushes it inhabits are studded with enormous fig-trees, to the fruit of which it is especially partial. It appears to have particular times in the day for feeding, and when thus engaged among the low shrub-like trees, I have approached within a few feet without creating alarm; but at other times the birds were extremely shy and watchful, especially the old males, which not unfrequently perch on the topmost branch or dead limb of the loftiest tree in the forest, whence they can survey all round, and watch the movements of their females and young in the brush below.

"In autumn they associate in small flocks, and may often be seen on the ground near the sides of rivers, particularly where the brush descends in a steep bank to the water's edge.

"The extraordinary bower-like structure first came under my notice in the Sydney Museum, to which an example had been presented by Charles Coxen, Esq., of Brisbane, as the work of the Satin Bower-bird. This so much interested me that I determined to leave no means untried for ascertaining every particular relating to this peculiar feature in the bird's economy; and on visiting the cedar-brushes of the Liverpool range, I discovered several of these bowers or playing-places on the ground, under the shelter of the branches of overhanging trees, in the most retired part of the forest: they differed considerably in size, some being a third larger than others. The base consists of an extensive and rather convex platform of sticks firmly interwoven, on the centre of which the bower itself is built: this, like the platform on which it is placed, and with which it is interwoven, is formed of sticks and twigs, but of a more slender and flexible description, the tips of the twigs so arranged as to curve inwards and nearly meet at the top : in the interior the materials are so placed that the forks of the twigs are always presented outwards, by which arrangement not the slightest obstruction is offered to the passage of the birds. The interest of this curious bower is much enhanced by the manner in which it is decorated with the most gaily-coloured articles that can be collected, such as the blue tail-feathers of the Rose-hill and Pennantian Parrakeets, bleached bones, the shells of snails, &c.; some of the feathers are inserted among the twigs, while others with the bones and shells are strewed about near the entrances. The propensity of these birds to fly off with any attractive object is so well known to the natives, that they always search the runs for any small missing article that may have been accidentally dropped in the brush. I myself once found at the entrance of one of them a small neatly-worked stone tomahawk, of an inch and a half in length, together with some slips of blue cotton rags, which the birds had doubtless picked up at a deserted encampment of the natives.

"It has been clearly ascertained that these curious bowers are merely sporting-places in which the sexes meet, and the males display their finery, and exhibit many remarkable actions; and so inherent is this habit, that the living examples, which have from time to time been sent to this country, continue it even in captivity. Those belonging to the Zoological Society have constructed their bowers, decorated and kept them in repair, for several successive years.

"In a letter received from the late F. Strange, he says—'My aviary is now tenanted by a pair of Satin-birds, which for the last two months have been constantly engaged in constructing bowers. Both sexes assist in their erection, but the male is the principal workman. At times the male will chase the female all over the aviary, then go to the bower, pick up a gay feather or a large leaf, utter a curious kind of note, set all his feathers erect, run round the bower, and become so excited that his eyes appear ready to start from his head, and he continues opening first one wing and then the other, uttering a low whistling note, and, like the domestic Cock, seems to be picking up something from the ground, until at last the female goes gently towards him, when, after two turns round her, he suddenly makes a dash, and the scene ends.'"

Mr. A. J. North writes to me:—"I forward you a photograph that may be of use to you in the preparation of your 'Monograph of the *Ptilonorhynchidæ, &c.*' It is that of a perfect bower of *Ptilonorhynchus violaceus* in the possession of the Trustees of the Australian Museum. It was found on the ground in the scrub near the Jenolan Caves, N. S. Wales, in December last, by Mr. J. C. Wiburd, and is built on a

platform of sticks and twigs about three inches in thickness, and is composed entirely of thin twigs slightly arched, some of which meet or cross each other at the top. Near the front of it, on the right side of the bower, is a tail-feather of *Platycercus elegans*. It measures over all 2 feet in length, 1 foot in height, and 10 inches in breadth; internally 8 inches in height by 4 inches in breadth. Scattered about the entrance are twelve pieces of bone of a small Wallaby (consisting of portions of the skull, ear-bones, lumbar vertebræ, and small bones of the feet), three pieces of moss, a spray of *Acacia* blossom, some small seed-cones of a *Eucalyptus*, an egg-bag of a spider, six specimens of a landshell (which my colleague Mr. Charles Hedley informs me is an unusual and remarkably keeled and depressed variety of *Thersites gulosa*, Gould), and one specimen of *Helicarion terrestris*. The photograph shows the front view of the bower only.

"These birds are at the present time committing great havoc in the orchards in the south coastal districts of the colony—probably from a scarcity of their normal food, owing to the late bush-fires and exceedingly dry season."

I have also received from Mr. Dudley Le Souëf the accompanying beautiful photograph of a bower of the present species found by him near Melbourne.

Dr. E. P. Ramsay describes the eggs of the Satin Bower-bird as follows:—

"The eggs vary in proportionate length, but are usually long ovals, seldom even slightly swollen towards the thicker end; the ground-colour is of a rich cream or light stone-colour, spotted and blotched with irregular patchy markings, and a few dots of amber and umber-brown of different tints, in some almost approaching blackish-brown, in others of a yellowish colour; the larger markings are, as usual, on the thicker end, but a few appear with the small dots on the thin end. In this, the usual form, the irregular short wavy lines previously mentioned seldom appear except where the larger spots or blotches are confluent; as if beneath the surface of the shell are a few irregularly shaped faint markings of slaty-grey or pale lilac. The eggs above described were taken from open nests composed of sticks and twigs, and lined with grass, by Mr. Ralph Hargrave, at Wattamolla, New South Wales."

The following descriptions are taken from my "Catalogue of Birds":—

Adult male. General colour above and below purplish black, the feathers having concealed greyish bases; upper tail-coverts black, broadly bordered and tipped with purple; quills and tail black, the feathers edged with purple: "bill bluish horn, passing into yellow at the tip; legs and feet yellowish white; iris beautiful light blue, with a circle of red round the pupil" (*Gould*). Total length 12.5 inches, culmen 1.4, wing 6.6, tail 4.5, tarsus 2.15.

Adult female. Different from the male. General colour above greyish green, with a shade of bluish on the edges of the feathers, the rump and upper tail-coverts greener than the back; lesser and median wing-coverts like the back, the latter edged with whity brown along the tips; greater and primary wing-coverts reddish brown, the innermost secondaries shaded with bluish and tipped with a bar of whity brown like the secondaries; tail-feathers golden brown, with a slight shade of bluish ashy on the centre feathers; lores and feathers round the eye a little browner than the head; ear-coverts and cheeks ashy brown, thickly streaked with yellowish-buff shaft-stripes; throat ashy brown, with a tinge of greenish grey, and slightly mottled with dusky greenish margins to the feathers; remainder of the under surface pale yellowish, the feathers all mottled with bars of blackish brown, tinged with bluish green, these bars less pronounced on the abdomen and under tail-coverts, the lower abdomen being uniform yellowish; axillaries pale greenish barred with dusky; under wing-coverts yellow, barred with dusky brown; quills dusky below, bright yellow at the base and on the inner web: "bill dark horn-colour; feet yellowish white, tinged with olive; irides of a deeper blue than in the male, and with only an indication of the red ring" (*Gould*). Total length 12 inches, culmen 1·3, wing 6·2, tail 4·6, tarsus 1·7.

Young male. Resembles the female at first, but is generally to be distinguished by a few purplish-black feathers appearing on the head and back or on the quills. The body-plumes appear to be acquired by a direct moult; but the quills and tail-feathers become black by a change in the colour of the feather itself.

The figures in the Plate represent an adult male and female of the size of life, with an illustration of a bower.

1. AMBLYORNIS FLAVIFRONS, *Rothschild.*
2.   ,,   INORNATA, *Schl. Adult Male.*

AMBLYORNIS FLAVIFRONS, *Rothsch.*

Yellow-fronted Gardener Bower-bird.

Amblyornis flavifrons, Rothschild, Novit. Zool. ii. p. 480 (1895).—Id. op. cit. iii. pl. i. figs. 3, 4 (1896).

This distinct species of yellow-crested Bower-birds was described by the Hon. Walter Rothschild in 1895 from a specimen in his collection from Dutch New Guinea. He has now two additional specimens in the Tring Museum, and there can be no doubt that it is distinct from the two other species of *Amblyornis*. The whole crest is more yellow than in *A. inornata* and *A. subalaris*, and this colour is continued down to the base of the bill, whereas in the other two species the forehead is brown like the back and the colour of the crest is orange. Mr. Rothschild further calls attention to the fact that the plumes of the crest in *A. flavifrons*, although very long and slender, have united webs like an ordinary feather, whereas in the other two species the webs are decomposed and each feather consists of a bundle of thin hair-like filaments. Again, as Mr. Rothschild observes, the colours of the underparts are distinctly separated at the chest in *A. flavifrons*, while in *A. inornata* the colour of the chest fades gradually towards the vent, and in *A. subalaris* the underparts are of a uniform brown, slightly spotted with buff.

The following is a description of the type-specimen in Mr. Rothschild's collection :—

General colour above dark brown, a little more rufescent on the lower back and rump; wing-coverts like the back; quills and tail dusky brown, externally washed with olive; crown of head from the base of the forehead bright orange-yellow, including the enormous crest; the shafts of the crest-feathers lemon-yellow towards the base; lores and sides of crown dark sooty brown, as well as the sides of the face and throat, shading off into lighter brown on the fore-neck and chest; remainder of under surface of the body light cinnamon-brown; axillaries cinnamon; under wing-coverts pale cinnamon; quills dusky below, yellowish along the inner web. Total length 8 inches, culmen 0·9, wing 5·2, tail 3·3, tarsus 1·3.

The lower figure represents the typical example of *A. flavifrons* of the natural size.

AMBLYORNIS INORNATA, *Schl.*

I have taken the present opportunity to give a figure of the male of this species, in full plumage, which has been discovered since the original Plate was drawn. For more than twenty years no yellow-crested bird had been received from the Arfak Mountains, the home of *A. inornata*; and I felt so convinced that the sexes were alike in colour, that I separated the southern form, *A. subalaris*, as a distinct genus, which I called *Xanthocklamys* (Bull. Brit. Orn. Club, iv. p. xv). It seems, however, that the male birds in collections received before 1894 must all have been young or not in full plumage, for when the adult male was discovered it turned out to have a magnificent orange crest, as was proved by Dr. Meyer from a specimen received by the Dresden Museum (Bull. Brit. Orn. Club, iv. p. xviii).

Since then fully adult males, in nuptial plumage, have been received by the Hon. Walter Rothschild not only from Arfak, but from the Owen Stanley Mountains in South-eastern New Guinea, where it has been discovered on Mt. Victoria and in the Rafa district between Mrs. Alexander and Bellamy. There can therefore be no doubt that the following synonyms also belong to *A. inornata* :—

Amblyornis musgravi, Goodwin, Proc. Zool. Soc. 1889, p. 451.

Amblyornis musgraveia, De Vis, Ann. Rep. Coll. Brit. New Guinea, p. 61 (1890).—Id. Colonial Papers, no. 103, p. 113 (1890).—Id. Ibis, 1891, p. 37.—Salvad. Ann. Mus. Civic. Genov. (2) x. p. 822 (1891).— Id. Aggiunte Orn. Pap. iii. p. 243 (1891).

Amblyornis musgravianus, Goodwin, Ibis, 1899, p. 553.

Xanthochlamys musgravianus, Sharpe, Bull. Brit. Orn. Club, iv. p. xiv (1894).

It should be noted that the form of the playing-ground as given by Mr. Goodwin is totally different from that sketched by Dr. Beccari.

AMBLYORNIS INORNATA (*Schlegel*).

Gardener Bower-bird.

Ptilonorhynchus inornatus, Schlegel, Nederl. Tijdschr. Dierk. iv. p. 51 (1871).—Rosenb. Reist. naar Geelvinkb
  pp. 102, 143 (1875).—Id. Malay Arch. pp. 554, 590 (1879).—Meunchenb. Dagboek, pp. 212, 249
  (1883).—Rosenb. MT. orn. Ver. Wien. 1885, p. 54.
Amblyornis inornata, Elliot, Ibis, 1872, p. 114.—Id. Meneng. Parad. pl. 37 (1873).—Sclater, P. Z. S. 1873, p. 697.
  —Salvad. Ann. Mus. Civic. Genov. vii. p. 761 (1875), ix. p. 193 (1876).—Beccari, Ann. Mus. Civic.
  Genov. ix. p. 382, tav. viii. (1877).—Id. Ibis, 1877, p. 395.—Salvad. op. cit. x. p. 161 (1877).—Gould,
  B. New Guin. i pl. 46 (1879).—D'Albertis, Nuova Guinea, p. 581 (1880).—Salvad. Orn. Papuasia e
  delle Molucche, i. p. 394 (1881).—Sharpe, Cat. Birds in Brit. Mus. vi. p. 394 (1881).—Salvad. Agg.
  Orn. Papuasia e delle Molucche, ii. p. 165 (1890).

WHEN this plain-looking Bower-bird was first discovered by Baron von Rosenberg, no one could have any idea of its peculiar talents for hut-building and garden decoration. It is now known, however, that two allied species exhibit the same curious habits in South-eastern New Guinea.

As far as has been recorded at present, the Gardener Bower-bird is an inhabitant of the Arfak Mountains in North-western New Guinea, and it has been procured in this region by all the best-known travellers who have visited this part of the globe; but for a detailed account of its "bower," science is indebted to the celebrated Italian naturalist, Dr. Beccari, who has published a description of it in the 'Annali' of the Genoa Museum. This was afterwards translated into the 'Gardeners' Chronicle' for March 6th, 1878, and the following passage is transcribed from the last-named journal:—

"The *Amblyornis inornata*, or, as I propose to name it, the Bird gardener, is a Bird of Paradise of the dimensions of a Turtledove. The specific name 'inornata' well suggests its very simple dress. It has none of the ornaments common to the members of its family, its feathers being of several shades of brown and showing no sexual differences.

"It was shot some years ago by the hunters of Mynheer von Rosenberg. The first descriptions of its powers of building (the constructions were called 'nests') were given by the hunters of Mynheer Bruijn. They endeavoured to bring one of the nests to Ternate; but it was found impossible to do this, both by reason of its great size and the difficulty of transporting it.

"I have fortunately been able to examine these constructions in the remote places where they are erected. On June 26, 1875, I left Andai for Hatam, on Mount Arfak. I had been forced to stay a day at Warmendi to give rest to my porters. At this time only five men were with me; some were suffering from fever, and the remaining porters declined to proceed. We had been on our way since early morning; and at 1 o'clock we intended to proceed to the village of Hatam, the end of our journey.

"We were on a projecting spur of Mount Arfak. The virgin forest was very beautiful. Scarcely a ray of sunshine penetrated the branches. The ground was almost destitute of vegetation. A little track-way proved that the inhabitants were at no great distance. A limpid fountain had evidently been frequented. I found here a new *Balanophora*, like a small orange or a small fungus. I was distracted by the songs and the screams of new birds; and every turn in the path showed me something new and surprising. I had just killed a small new marsupial (*Phascologale dorsalis*, Pet. and Doria) that balanced itself on the stem of a great tree like a squirrel; and turning round, I suddenly stood before a most remarkable specimen of the industry of an animal. It was a hut or bower close to a small meadow enamelled with flowers. The whole was on a diminutive scale. I immediately recognized the famous nests described by the hunters of Bruijn. I did not, however, then suspect that they had anything to do with the constructions of the *Chlamydoderæ*. After well observing the whole, I gave strict orders to my hunters not to destroy the little building. That, however, was an unnecessary caution, since the Papuans take great care never to disturb these nests or bowers, even if they are in their way. The birds had evidently enjoyed the greatest quiet until it happened, unfortunately for them, to come near them. We had reached the height of about 4800 feet; and after half an hour's walk we were at our journey's end.

"*The Nest.*—I had now full employment in the preparation of my treasure; and I gave orders to my people not to shoot many of the birds. The nest I had seen first was the nearest one to my halting-place. One morning I took colours, brushes, pencils, and gun, and went to the spot. I there made the sketch which I now publish. While I was there neither host nor hostess was at home. I could not wait for them. My hunters saw them entering and going out, when they watched their movements to shoot them. I could not ascertain whether this bower was occupied by one pair or by several pairs of birds, or whether the sexes were in equal or unequal numbers—whether the male alone was the builder, or whether the wife assisted in the construction. I believe, however, that such a nest lasts for several seasons.

"The *Amblyornis* selects a flat even place around the trunk of a small tree that is as thick and as high as a walking-stick of middle size. It begins by constructing at the base of the tree a kind of cone, chiefly of moss, of the size of a man's hand. The trunk of the tree becomes the central pillar; and the whole building is supported by it. On the top of the central pillar twigs are then methodically placed in a radiating manner, resting on the ground, leaving an aperture for the entrance. Thus is obtained a conical and very regular hut. When the work is complete many other branches are placed transversely in various ways, to make the whole quite firm and impenetrable. A circular gallery is left between the walls and the central cone. The whole is nearly 3 feet in diameter. All the stems used by the *Amblyornis* are the thin stems of an orchid (*Dendrobium*), an epiphyte forming large tufts on the mossy branches of great trees, easily bent like straw, and generally about 20 inches long. The stalks had the leaves, which are small and straight, still fresh and living on them—which leads me to conclude that this plant was selected by the bird to prevent rotting and mould in the building, since it keeps alive for a long time, as is so often the case with epiphytical orchids.

"The refined sense of the bird is not satisfied with building a hut. It is wonderful to find that it has the same ideas as a man; that is to say, what pleases the one gratifies the other. The passion for flowers and gardens is a sign of good taste and refinement. I discovered, however, that the inhabitants of Arfak did not follow the example of the *Amblyornis*. Their houses are quite inaccessible from dirt.

"*The Garden.*—Now let me describe the garden of the *Amblyornis*. Before the cottage there is a meadow of moss. This is brought to the spot and kept free from grass, stones, or anything which would offend the eye. On this green tuft flowers and fruits of pretty colour are placed so as to form an elegant little garden. The greater part of the decoration is collected round the entrance to the nest; and it would appear that the husband offers there his daily gifts to his wife. The objects are very various, but always of vivid colour. There were some fruits of a *Garcinia* like a small-sized apple. Others were the fruits of *Gardenias* of a deep yellow colour in the interior. I saw also small rosy fruits, probably of a Scitamineous plant, and beautiful rosy flowers of a splendid new Vaccinium (*Agapetes amblyornidis*). There were also fungi and mottled insects placed on the turf. As soon as the objects are faded they are moved to the back of the hut.

"The good taste of the *Amblyornis* is not only proved by the nice house it builds. It is a clever bird, called by the inhabitants 'Buruk Gurea' (master bird), since it imitates the songs and screamings of numerous birds so well that it brought my hunters to despair, who were but too often misled by the *Amblyornis*. Another name of the bird is 'Tukan Kobon,' which means a gardener."

When Mr. D. G. Elliot founded the genus *Amblyornis*, he separated it on account of the more exposed nostrils and from its having ten tail-feathers instead of twelve. When he wrote, only one specimen was known, and it has since turned out that *Amblyornis* has really twelve tail-feathers, two being deficient in the original specimen. The difference in the feathering over the nostrils is only one of degree, but, as Count Salvadori has pointed out, there are other good characters, such as the shape of the bill, with its smooth operium, and the want of scutellations on the tarsus, which distinguish *Amblyornis*. One of the most striking of the characters in the genus is the similarity in colour of the sexes.

Adult. General colour above brown, rather more reddish on the head and mantle; wing-coverts like the back; quills and tail dusky brown, externally like the back; lores and sides of face dull brown; throat and under surface of body orange-brown, rather lighter on the abdomen; sides of breast washed with the same brown as the back; axillaries and under wing-coverts brighter orange-buff; quills light brown below, pale buff along the edge of the inner web: "bill black; feet pale lead-colour; iris chestnut" (*D'Albertis*). Total length 9·5 inches, culmen 1·1, wing 5·05, tail 3·4, tarsus 1·4.

The female is similar to the male in colour.

The description and figure are taken from a specimen in the British Museum, formerly in the Gould Collection. The "bower" is drawn from the materials published by Dr. Beccari.

AMBLYORNIS SUBALARIS, *Sharpe.*

Orange-crested Bower-bird.

Amblyornis subalaris, Sharpe, Journ. Linn. Soc., Zool. xvi. p. 408 (1884).—Finsch v. Meyer, Zeitschr. ges. Orn. ii. p. 390, pl. xxi. (1885).—Id. Ibis, 1886, p. 257.—Sharpe, Nature, p. cccxi (1886).—Id. in Gould's B. New Guin. vol. i. pl. 47 (1886).—D'Hamonv. Bull. Soc. Zool. France, xi. p. 511 (1886).—Ramsay, Proc. Linn. Soc. N. S. W. (2) ii. p. 250 (1887).—Salvad. Agg. Orn. Pap. ii. p. 165 (1890).—Goodwin, P. Z. S. 1889, p. 451.—Id. Ibis, 1890, p. 155.—De Vis, Ann. Rep. Brit. New Guinea, p. 51 (1890).—Id. Colonial Papers, no. 103, p. 112 (1890).—Id. Ibis, 1891, p. 37.—Salvad. Agg. Orn. Pap. iii. p. 243 (1891).—De Vis, Ann. Rep. Brit. New Guinea, 1890–91, p. 95 (1892).

This remarkable species of Bower-bird was first discovered by Mr. Goldie, in the Astrolabe Range of the Owen-Stanley Mountains, in South-eastern New Guinea. He procured only the female bird, which remained in the British Museum for many months before I ventured to describe it as distinct from the Bower-bird of the Arfak Mountains, *A. inornata*. It seemed to be, however, a distinct species, and I at last gave it a name, little dreaming that in the following year the male bird would be discovered, and would turn out to be such a beautiful and striking form of Bower-bird. Although resembling *A. inornata* in the plain brown plumage of the body, it excels that species in the possession of a gorgeous orange crest. This led me to suppose that *A. inornata* might also be found to possess an equally brilliant decoration of the head, but Count Salvadori aptly remarks that too many specimens of the Arfak bird, of both sexes, have been received by European Museums to render it possible that the species possesses any crest or particular ornament, such as we find in the Bower-bird from South-eastern New Guinea.

The adult males referred to above were procured by the late Mr. Carl Hunstein in that part of the Astrolabe Mountains which he called the Horseshoe Range, and Sir William Macgregor has also discovered the species on Mount Musgrave, at a height of from 6000 to 9000 feet, and on Mount Suckling, at a height of 4100 feet. He procured a male bird in the vicinity of its bower, which is described in Mr. De Vis's Zoological Report attached to the Blue-book of 1892:—

"This bower is built of twigs arranged in the shape of a shallow circular basin, about 3 feet in diameter, the side being some 6 inches higher than the centre. The whole of the basin is covered with a carpet of the greenest and most delicate moss, which, as it is of a different kind to that growing around on the ground, trees, roots, &c., led me to conjecture that it had been planted by the bird itself. The surface is scrupulously cleared of all leaves, twigs, &c. In the centre of the basin a small tree, without branches, about 2 inches in diameter, is growing. Immediately around this tree, and supported by it to the height of about 2 feet, is erected a light structure of small sticks and twigs, placed horizontally, and crossing one another. On the extreme outer edge of the basin a more substantial collection of twigs had been built up, which was arched above so as to join the collection around the centre pole, leaving a clear space beneath for the bird to pass through in his gambols. The basin has two entrances leading into it. They are 4 or 5 inches apart, and are formed by a depression or gap in the outer rim. The bower is placed immediately to the right of the entrances. At the opposite side to the entrances, and on the highest part of the raised rim of the basin, is placed a quantity of black sticks (4 inches or so in length), black berries, and the black wing-coverings of large coleoptera. Black is evidently the most attractive colour to this bird."

I have reproduced the illustrations which accompany Mr. De Vis's Report.

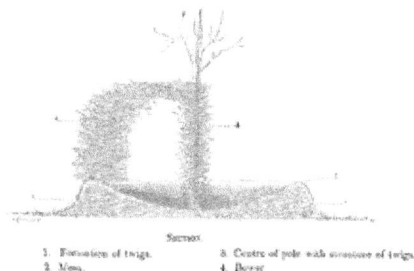

Screen.

1. Formation of twigs.   3. Centre of pole with structure of twigs.
2. Moss.   4. Bower.

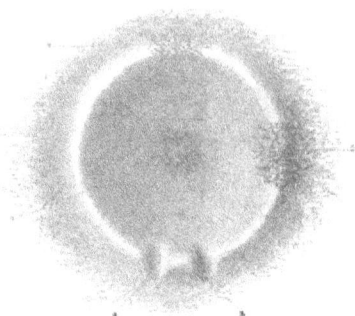

GROUND-PLAN.

1. Centre path.   3. Entrances.
2. Bower.   4. Twigs, berries, and beetles.

Mr. Goodwin's account of the bower of the present bird differs somewhat from that given above.

The following is a description of a pair of adult birds :—

Adult male. General colour above uniform dark olive-brown, rather more olive on the back, rump, and upper tail-coverts ; wing-coverts like the back ; bastard wing, primary-coverts, and quills olive-brown externally, internally dark brown ; tail-feathers dark brown, washed with olive-brown externally ; crown of head with an immense crest of orange, the lateral and frontal feathers edged and tipped with blackish brown ; base of forehead dusky olive-brown ; hind neck lighter olive-brown ; lores ashy ; sides of face, eyebrow, and ear-coverts dark olive-brown ; cheeks and entire under surface of body light olive-brown, streaked down the centre of the feathers with ochreous buff, the sides of body and flanks rather browner ; thighs dusky brown ; under tail-coverts fulvous, with ochreous-buff centres to the feathers, the long ones edged with dark brown ; under wing-coverts and axillaries orange-buff or tawny ; quills below dusky, ochreous along the inner edge. Total length 8·3 inches, culmen 1, wing 5, tail 3·4, tarsus 1·3.

Adult female. Differs from the male in having no orange crest, the head being like the back. Total length 8·3 inches, culmen 0·9, wing 4·8, tail 3·3, tarsus 1·4.

Mr. Forbes procured specimens of both sexes, killed in the rainy season. The whole of the colours are paler and more olive, and the ochreous tints of the under surface are much paler, especially on the under wing-coverts. The male is distinguished from the female at this season of the year only by the greater amount of clear ochreous on the underparts.

The figures in the Plate represent an adult male and female, drawn from a pair procured by Mr. Hunstein in the Horseshoe Range.

CNEMOPHILUS MACGREGORII, De Vis.

Macgregor's Bird of Paradise.

Xanthomelus macgregori, Goodwin, Ibis, 1890, p. 150.

Cnemophilus macgregorii, De Vis, Ann. Rep. Brit. New Guinea, p. 61 (1890).—Id. Colon. Papers, no. 103, p. 115 (1890).—Id. Ibis, 1891, p. 49.—Sclater, Ibis, 1891, p. 414, pl. x.

This remarkable form was discovered by Sir William Macgregor during his expedition to the Owen Stanley Mountains, and was procured at Mount Knutsford, at an elevation of 11,000 feet. The only specimen as yet known is an adult male, which is at present in the Queensland Museum, but the courteous Director, Mr. C. W. De Vis, sent it over to Europe to Dr. Sclater, who described and figured it in 'The Ibis.'

Mr. Goodwin, who visited England shortly after the close of the Macgregor Expedition, to which he was attached as one of the naturalists, communicated an account of the Birds of Paradise observed by him to 'The Ibis,' and gave a description of this species from memory, which is characterized by Count Salvadori as a "descriptio incompleta." Mr. De Vis, however, to whom was intrusted the description of the natural-history objects obtained by the expedition, gave a very full description of the species, for which he created the name of *Cnemophilus*. That he was right in placing it in a distinct genus is beyond question. Dr. Sclater, in his paper on the species, has so well summarized its characters that I cannot do better than quote his remarks:—

"There is certainly a general resemblance in colour and shape between *Cnemophilus* and *Xanthomelus*, and the feet in both forms are large and strong, although this feature is carried to a much greater extent in *Xanthomelus*, which has the tarsi much stronger and rather longer than *Cnemophilus*. In *Xanthomelus*, moreover, the scutellations of the front of the tarsus are well marked, whereas in *Cnemophilus* the scutella are fused into one nearly uniform plate. The wings of *Cnemophilus* are much shorter and more rounded than those of *Xanthomelus*. But it is in the bill of these two forms that the greatest divergence is observable.

"In *Xanthomelus* the bill is long and strong, the loral plumes are short, and the base of the bill, nostrils, and culminal ridge are quite bare. In *Cnemophilus* the bill is shorter and not so thick, the loral plumes are elongated, projecting forwards, and covering the base of the bill so far as to partially cover the nostrils. Besides this the frontal plumes are elongated and elevated into a compressed ridge, which is carried forward over the culmen and backward to the base of the very singular thin crest, composed of five or six lengthened feathers, which spring up immediately behind the front.

"In these last characters *Cnemophilus* is quite distinct from other birds, but obviously approaches *Diphyllodes*. I should be disposed, therefore, to place *Cnemophilus* along with the Paradise-birds rather than along with the Bower-birds, if these two groups are to be kept apart. But there can be no doubt that the Bower-birds are closely allied to the Paradise-birds, and several well-known recent authorities have united them into one family."

The figure in the Plate has been drawn from a picture painted by Mr. Keulemans from the type specimen which Mr. De Vis so kindly sent to England for examination.

PRIONIDURA NEWTONIANA, *De Vis.*

PRIONODURA NEWTONIANA, De Vis.

Newton's Bower-bird.

Prionodura newtoniana, De Vis, Proc. Linn. Soc. N. S. Wales, vii. p. 562 (1883).—Ramsay, Tabular List of Austr. B. p. 11 (1888).—De Vis, Proc. Roy. Soc. Queensl. 1889, p. 245.—id. Rep. Exped. Bellenden-Ker Range, p. 87 (1889).—Meston, t. c. p. 120.—Sclater, Ibis, 1890, p. 264.

Corymbæcola mestoni, De Vis, MSS.; *fide* Meston, l. c.

This remarkable Bower-bird was described by Mr. De Vis from a single specimen procured in Queensland by Mr. Kendal Broadbent, and was at once recognized by its describer as belonging to a new genus. As with my *Amblyornis subalaris*, it has since transpired that the type specimen of *Prionodura* was a female or young male, the bird in neither case giving promise of having a brilliantly decorated adult male. The genus is closely allied to *Amblyornis*, but differs in the style of ornamentation in the male.

In a "Further Account of *Prionodura newtoniana*," Mr. De Vis has given the subsequent history of the species, and I cannot do better than reproduce his own words:—

"The bird was first discovered by Mr. K. Broadbent in the scrubs clothing the banks of the Tully River, a small river issuing from an angle formed by spurs of the Coast Range on its eastern aspect and entering the sea some little distance to the north of Cardwell. In the vale of the Herbert, on the western side of the principal spur and more immediately in the vicinity of Cardwell, the bird does not seem to occur, Mr. Broadbent having there searched for it more than once without success—lat. 18° is, therefore, probably its southern limit of range. Its true habitat is now ascertained to be the highlands north of the township of Herberton, where it was first observed by Mr. A. Meston in the course of a flying visit to the top of Bellenden-Ker. From near the summit of this mountain Mr. Meston brought down the skin of a male bird; and soon after, Mr. Broadbent, visiting Herberton in pursuit of the Tree-Kangaroo of that district, encountered the bird frequently about seven miles from town (fifty miles from the Bellenden-Ker), and collected a rich series of examples. How far northward the bird extends its range is as yet unknown.

"*Prionodura* is emphatically a Bower-bird. Both its observers in nature met with its bowers repeatedly and agree in representing them to be of unusual size and structure. From their notes and sketches it would appear that the bower is usually built on the ground between two trees, or between a tree and a bush. It is constructed of small sticks and twigs. These are piled up almost horizontally around one of the trees in the form of a pyramid, which rises to a height varying from four to six feet; a similar pile of inferior height, about eighteen inches, is then built round the foot of the other tree; the intervening space is arched over with stems of climbing plants, the piles are decorated with white moss, and the arch with similar stems mingled with clusters of green fruit resembling wild grapes. Through and over the covered run play the birds, young and old, of both sexes. A still more interesting and characteristic feature in the play-ground of this bird remains. The completion of the massive bower so laboriously attained is not sufficient to arrest the architectural impulse. Scattered immediately around are a number of dwarf hut-like structures—'gunyahs,' they are called by Broadbent, who says he found five of them in a space of ten feet diameter and observes that they give the spot exactly the appearance of a miniature blacks' camp. These seem to be built by bending towards each other strong stems of standing grass and capping them with a horizontal thatch of light twigs. In and out and around the 'gunyahs,' and from one to another, the birds in their play pursue each other to their hearts' content."

Mr. De Vis gives further notes in the Report of the Bellenden-Ker expedition:—"Found at all heights to the summit of Bellenden-Ker and in the scrubs around Herberton at a high elevation. In connection with the bower of this handsome bird we are indebted for an interesting fact to Broadbent's observation, that whereas towards the base of the mountain the bowers have the elaborate formation noticed lately in the Proceedings of the Royal Society of Queensland, at higher levels they gradually lose their distinctive character, and at the top are reduced to the simple trough-like form of the bower of the Regent- and Satin-birds, for which they might be mistaken were those birds inhabitants of the district. There is reason to believe that Mr. Meston has acquainted us with the nest and egg of this bird. While hearkening to the call of a male he noticed a rustling in a bush by his side, and looking in saw a bird which he says, without hesitation, was the female just disturbed from a nest built in a fork of the bush. The nest in question is cup-shaped and loosely constructed of fibrous roots, lined with finer material of the same kind, and decorated with a little green moss on the outer side. The egg is 27 mm. long, 9 mm. broad, pale yellowish-grey, profusely freckled and blotched with pale brown."

The following account of the habits of the species is given by Mr. Meston in the "Report of the Government Scientific Expedition to Bellenden-Ker Range." It will be noticed that he speaks of the species as Meston's Bower-bird; but the bird is named as a compliment to Professor Alfred Newton; and without detracting from the merit of Mr. Meston's having shot the first adult male, it is obvious that the English name of this species must follow the Latin designation and be associated with the celebrated naturalist in whose honour the name of *newtoniana* was given.

"Most remarkable of all the birds named by De Vis is *Prionodura newtoniana*, or 'Meston's Bower-bird.' The name requires some explanation. On my first ascent of the mountain I shot a full-plumaged male specimen at 4800 feet. This was regarded by De Vis as one of an entirely new species and named *Corythaila meston*. Subsequently it appeared that a young uncoloured male had been previously shot by Broadbent on the head of the Tully, and received from De Vis the name of *Prionodura newtoniana*. To me, therefore, belonged simply the honour of having shot the first full-plumaged male and observed the habits of this extraordinary bird, and the final name, to be known hereafter to science, is *Prionodura newtoniana*, or Meston's Bower-bird. Since the first male was found by me, several males and females have been shot by Broadbent on the Herberton Ranges at 3500 feet. The blacks on the Mulgrave and Russell call this bird 'Warpundilla.' So far it is unknown south of the Tully or north of the Barron. During the expedition we obtained seven males in perfect plumage and several females. This is one of the three handsomest birds in Australia, the other two being the Rifle-bird and Regent-bird—*Ptiloris victoriæ* and *Sericulus melinus*. In habits and peculiarities it is one of the most eccentric birds in the world. The lowest descent was 1500 feet, between the summit of Bernard's Spur and the Wheelaman Pools. Usually it is found from 4800 feet to 5000 feet. The note of the female—a bird of common grey plumage—is that of the ordinary green Cat-bird, in a sharper and shriller key. The male appears to possess the marvellous imitative powers of the Australian Lyre-bird. First you hear him croaking like a Tree-Frog, and this note is followed by a low, soft, musical, pathetic whistle, succeeded in a rapid succession by an astonishing imitation of apparently all the birds in the scrub. The bower varies in size and shape, but in all cases differs from that of the other Australian Bower-birds. Both Broadbent and myself have seen bowers up to a height of 8 feet. As a rule, they are made between two small trees about 4 feet or 5 feet apart. Short dead sticks and twigs are piled up against each tree in a gently contracting pyramid, and across from base to base extends an arch-shaped causeway, occasionally spanned by a connecting vine, decorated with green mosses and tufts of tiny ferns. In and out and over and under and around this erratic structure both male and female birds disport themselves in frequent playful festivities, like the Lyre-bird, Regent-bird, and Satin- and other Bower-birds remarkable for similar customs and proclivities. So far only one nest has been discovered—the one found by me on the summit of the Little Mulgrave Range. It was a round cup-shaped nest, decorated outwardly with the mosses and ferns used in ornamenting the bowers. It contained only one egg, quite fresh; so we have yet to learn if the bird lays one or more."

The following descriptions are taken from a pair of birds in the British Museum, collected by Mr. Kendal Broadbent on the Bellenden-Ker mountain:—

Adult male. General colour above golden-olive, slightly brighter on the rump and upper tail-coverts; wing-coverts and quills golden-olive, the inner webs dingy brown, with a broad margin of pale yellow on the inner web; the innermost secondaries dusky olive on the inner web; the centre tail-feathers olive-brown, with a golden wash, the extreme base of the feathers bright yellow; the next two feathers bright yellow, with a broad tip of olive-brown; the succeeding feather with a narrower tip of olive-brown, and the three outer tail-feathers entirely bright yellow; crown of head olive-brown with a golden wash, the entire sides of the face of this same colour; the crown with a broad median crest of golden-yellow; the nape and hind neck also golden-yellow, this colour overspreading the upper mantle; cheeks and chin olive-brown like the ear-coverts; the whole of the remainder of the under surface bright golden-yellow, with a slight wash of golden-olive on the flanks; thigh-feathers ashy, tipped with yellow; under tail-coverts deep golden-yellow like the under surface of the tail; axillaries and under wing-coverts golden-yellow; quill-lining also yellow, as well as the shafts of the feathers underneath. Total length 9·5 inches, culmen 0·9, wing 4·85, tail 4·2, tarsus 1·2.

Adult female. Different from the male. Entirely olive-brown above, with ashy shaft-streaks to the feathers of the head and neck; wing-coverts like the back; the bastard-wing, greater coverts, and outer aspect of quills a little browner than the back; tail-feathers brown, washed slightly with olive near the base of the outer webs; lores ashy grey; ear-coverts and cheeks ashy grey, washed with olive-brown; under surface of body pale ashy grey, becoming lighter on the abdomen and under tail-coverts, the feathers on the lower throat and breast with whitish shaft-lines; axillaries and under wing-coverts ashy grey, the lower ones edged with yellowish; the quills dusky brown below, yellow along the basal two-thirds of the inner web. Total length 9 inches, culmen 0·85, wing 4·5, tail 3·35, tarsus 1·2.

The Plate represents the male and female of the size of life; the figures being drawn from the pair of birds described above.

XANTHOMELUS AUREUS.

XANTHOMELUS AUREUS (*Linn.*).

Golden Bird of Paradise.

Golden Bird of Paradise, Edwards, Nat. Hist. Birds, iii. p. 112, pl. 112 (1750).—Lath. Gen. Syn. i. pt. 2, p. 453 (1782).

Oriolus aureus, Linn. Syst. Nat. i. p. 163 (1766).—Gm. Syst. Nat. i. p. 394 (1788).—Vieill. N. D. d'Hist. Nat. xviii. p. 194 (1817).—Temm. Pl. Col. ii. Genre Loriot (1825).—Wagl. Syst. Av., *Oriolus* sp. (1827).—Gray, Gen. B. ii. p. 232 (1842).—Id. Cat. B. New Guin. p. 57 (1856).—Id. Hand-l. B. i. p. 293, no. 4332 (1869).—Rosenb. Reist. naar Geelvinkb. p. 117 (1875).—Mussehenbr. Dagboek, pp. 209, 236 (1883).—Rosenb. Mitth. orn. Ver. Wien, 1883, p. 54.

Le Rollier de Paradis, Montb. Hist. Nat. Ois. iii. p. 149 (1775).

Paradisea aurea, Bocwnsk. Nat. ii. p. 122 (1780-84).—Lath. Ind. Orn. i. p. 197 (1790).—Cuv. Règn. An. i. p. 405 (1817).—Swains. Zool. Journ. i. p. 472, note (1825).—Schleg. J. f. O. 1861, p. 385.

Le Paradis orangé, Vieill. Ois. Dor. ii. Ois. Parad. p. 26, pls. 11, 12 (1802).

Le Loriot de Paradis, Levaill. Ois. de Parad. i. pls. 18, 19 (1806).

Paradisea aurantia, Shaw, Gen. Zool. vii. p. 492, pl. 68 (1809).

Lophorina aurantia, Steph. Gen. Zool. xiv. p. 76 (1826).

Sericulus aureus, Less. Man. d'Orn. i. p. 396 (1828).—Bp. Consp. Av. i. p. 349 (1850).—Gray, P. Z. S. 1858, p. 192, 1861, p. 434.—Wall. Ibis, 1861, p. 287.—Id. P. Z. S. 1862, pp. 154, 157, 159, 160.—Rosenb. Nat. Tijdschr. Ned. Ind. xxv. p. 236 (1863).—Id. J. f. O. 1864, p. 120.—Finsch, Neu-Guinea, p. 173 (1865).—Schleg. Mus. P.-B., Coraces, p. 98 (1867).—Wall. Malay Archip. ii. pp. 419, 420 (1869).—Beccari, Ann. Mus. Genov. vii. p. 709 (1875).—D'Albertis, t. c. p. 738.—Scl. Ibis, 1876, p. 248.—Rosenb. Malay. Archip. p. 554 (1879).

Oriolus aurantiacus, Dumont, Dict. Class. Sc. Nat. xxvii. p. 216

Loriot de Paradis orangé, Less. Voy. Coq., Zool. i. pt. 2, p. 654 (1828).

Sericulus aurantiacus, Less. Tr. d'Orn. p. 339 (1831).—Id. Ois. Parad. Syn. p. 20, et Nat. Hist. p. 201, pl. 25 (1835).—Id. Compl. de Buff., Ois. p. 497 (1838).

Xanthomelus aureus, Bp. Compt. Rend. xxxviii pp. 262, 538 (1854).—Scl. Journ. Pr. Linn. Soc. ii. p. 160 (1858).—Elliot, Ibis, 1872, p. 112.—Id. Monogr. Parad. pl. xv. (1873).—Salvad. Ann. Mus. Civ. Genov. vii. p. 783 (1875), ix. p. 192 (1876).—Id. Atti R. Ac. Sc. Tor. xi. p. 658 (1876).—Id. Ibis, 1876, p. 267.—Beccari, Ann. Mus. Civ. Genov. ix. p. 385 (1877).—Salvad. op. cit. x. p. 14 (note), p. 152 (1877).—Sharpe, Cat. B. Brit. Mus. iii. p. 186 (1877).—Gould, B. New Guin. i. pl. 48 (1875).—D'Alb. Nuova Guin. pp. 28, 80, 83, 582 (1880).—Études-Deslongch. Ann. Mus. d'Hist. Nat. Caen, i. p. 39 (1880).—Salvad. Orn. Papuasia, ii. p. 568 (1881).—Guillem. P. Z. S. 1885, p. 557.—D'Hamonv. Bull. Soc. Zool. Fr. 1886, pp. 508, 511.—Salvad. Agg. Orn. Papuasia, ii. p. 164 (1890).

Oriolus xanthogaster, Rosenb. in Isis (1871).—Id. Reist. naar Geelvinkb. pp. 117, 139 (1875).—Id. Mitth. orn. Ver. Wien, 1883, p. 54.

Sericulus xanthogaster, Schleg. Ned. Tijdschr. Dierk. iv. p. 50 (1871, *ex* Rosenb.).—Rosenb. Reist. naar Geelvinkb. p. 102 (1875).—Salvad. Atti R. Ac. Sc. Tor. xi. p. 688 (1876).—Id. Ibis, 1876, p. 267.—Rosenb. Malay. Archip. pp. 554, 599 (1879).

Chlamydodera xanthogaster, Elliot, Ibis, 1872, p. 113.—Id. Monogr. Parad. pl. 33 (1873).—Scl. P. Z. S. 1873 p. 697.

THE line of demarcation between the Birds of Paradise and the Bower-birds has never been easy of definition, and some naturalists, like Mr. D. G. Elliot and Count Salvadori, have united the two families together. As the nesting-habits and bower-building propensities of the two groups are gradually being discovered, it would seem that many of the species hitherto considered to be true Birds of Paradise are also the frequenters of "playing-grounds," after the manner of the Bower-birds.

The present species has generally been placed at the end of the series of Birds of Paradise, as being apparently the nearest ally of the Bower-birds of the genus *Amblyornis*, and the coloration of the females and young birds so closely resembles that of a Bower-bird that there is nothing remarkable in the fact that, when they were first discovered by Baron von Rosenberg, they were described by Professor Schlegel as *Sericulus xanthogaster* and afterwards placed by Mr. Elliot in the genus *Chlamydodera*.

The nesting-habits of the Golden Bird of Paradise are not yet exactly known. Dr. Beccari was informed by the natives that it nested on the ground, but he was not able to discover whether it made a bower like *Chlamydodera*, or built a hut like *Amblyornis*. Count Salvadori has, however, expressed his opinion that the *Xanthomeli* will be found to construct a bower of some sort, and of this I myself likewise entertain no doubt. As far as is known at present, the Golden Bird of Paradise is only found in the north-western portion of New Guinea, being replaced by *Xanthomelus ardens* in the south-eastern part of the island. Its supposed occurrence in the island of Waigiou is, as pointed out by Count Salvadori, a mistake; and the only localities inhabited by the species, of which we have exact information, are Dorey, where Lesson and the Dutch travellers procured it, the Arfak Mountains, where D'Albertis and Beccari met with it, and Sorong, where it was also found by D'Albertis. Mr. Wallace procured a native-prepared skin in Salawati, but this may have been brought from the mainland of New Guinea, as no living specimens have been observed in that island.

The only notes on the habits of the species are those given by Dr. Beccari :—" It was procured by me near Hatam, on the same fig-tree on which D'Albertis obtained the greater number of his birds. It has more or less the habits of a Bird of Paradise, feeding on fruits and particularly on figs. Not more than two or three individuals are found together, generally only a male and a female. It is a very lively and suspicious bird, and after I had killed a male bird, a female, accompanied by another bird (probably a young one), came back after a while to feed on the same tree, but I could not discover them. Although this bird is found up to an altitude of 3000 feet and more, it seems to be more abundant on the hills near the sea. It is always difficult to find, and even in the places which it frequents there never seem to be more than two pairs. The note, according to my hunters, resembled the *zigodo* of the Sun-birds, but was much stronger and louder. Only the tuft of plumes of the head is erectile. By the natives of Arfak it is called *Komisk.*"

The description of the male is taken from the ' Catalogue of Birds,' and that of the female and young bird from Count Salvadori's ' Uccelli di Papuasia.'

"*Adult male.* General colour above fiery orange-red, the head crested, the back with a large dorsal shield; scapulars, as well as the lower back, rump, and tail-coverts, orange-yellow mixed with black at the base; wings deep orange, the feathers black at the base; quills orange, shading off into olive-brown, with more or less orange-yellow towards the base; tail black; a narrow line running from the base of the bill, as well as the lores and the feathers round the eye, sides of face, and throat, black; rest of under surface of body orange-yellow; sides of head and neck, as well as the long plumes on the side of the latter, fiery orange-red like the dorsal shield. Total length about 9 inches, culmen 0·95, wing 5·1, tail 3·65."

"*Adult female.* Above dusky olivaceous; sides of head and throat dusky black, varied with longitudinal spots of olive; breast, abdomen, and under tail-coverts yellow." (*Lesson.*)

"*Young.* Above brown, with scarcely any olive; the shafts of the concealed portion of the interscapulary plumes yellow; under surface of body bright yellow; throat and sides of head pale brownish rufous; upper breast ornamented with dusky angulated bands; wings and upper surface of body coloured like the back, and scarcely any darker; under wing-coverts and basal part of the inner web of the quills yellow; lower surface of the shafts of the quills yellow; tail olive-yellow underneath; bill dusky black; feet leaden grey; iris chestnut."

The Plate is the same as that published in the late Mr. Gould's ' Birds of New Guinea.'

XANTHOMELUS ARDENS, *D.Albert. & Salvad.*

W. Hart del. et lith.   Mintern Bros. imp.

XANTHOMELUS ARDENS, *D'Alb. & Salvad.*

Southern Golden Bird of Paradise.

Sericulus aureus (pt.), D'Alb. Ann. Mus. Civ. Gen. vii. p. 798 (1875).—Id op. cit. x. pp. 14, 29 (1877).

Xanthomelus aureus?, D'Alb. & Salvad. Ann. Mus. Civ. Gen. xiv. p. 113 (1879).—D'Alb. Nuova Guinea, pp. 241, 388, 453, 580 (1880).

Xanthomelus ardens, D'Alb. & Salvad. Ann. Civ. Gen. xiv. p. 113 (1879).—Salvad. Orn. Papuasia, etc. ii. p. 665 (1881).—Id. Aggiunte Orn. Papuasia, etc. ii. p. 165 (1890).—Sharpe, Bull. B. O. C. iv. p. xiv (1894).

Oriolus ardens, Meurschmbr. Dagboek, pp. 216, 237 (1883).—Reichb. Mitth. orn. Ver Wien, 1885, p. 54.

This species, which is a southern representative of *X. aureus*, was discovered by Signor D'Albertis on the Fly River, where he procured a native skin of a male and also a perfect skin of a young bird. He says that the plumes of this bird are worn by the natives of the Fly River as ornaments. No one appears to have met with the species since the visit of D'Albertis to this part of New Guinea.

Adult male. Similar to *X. aureus*, but of a much more brilliant and fiery red on the head, neck, and mantle.

Young male. Above brown, washed with ashy grey on the edges of the feathers of the scapulars and mantle, all of which have yellow shafts; quills brown, externally yellowish brown or golden olive, some of the wing-coverts and scapulars washed externally with the latter colour; tail feathers washed with ashy along the outer webs; head and neck lighter brown than the back, the feathers on the sides of the neck longer and forming a frill; sides of face and ear-coverts light brown, the latter rufescent; chin isabelline, fading off into the yellow of the throat, which is pale and forms a narrow band shut in by the frilled sides of the neck; all the rest of the under surface of the body bright golden yellow, paler on the thighs; sides of upper breast slightly washed with light brown; under wing-coverts, axillaries, and quill-lining also bright golden yellow : " bill reddish brown, yellow at base of lower mandible, feet olivaceous besides grey; irides yellow " (*D'Albertis*). Total length 10 inches, culmen 1·1, wing 5·4, tail 2·8, tarsus 1·75.

The figures and descriptions given are taken from the type specimens kindly lent me by the Marquis Doria.

SERICULUS MELINUS (Lath.)

SERICULUS MELINUS (Lath.).

Regent Bird.

Yellow-bellied Thrush, Lath. Gen. Syn. Suppl. ii. p. 187 (1801).

Turdus melinus, Lath. Ind. Orn. Suppl. ii. p. xlv (1801).—Vieill. N. Dict. d'Hist. Nat. xx. p. 243 (1818).—
Id. Enc. Méth. iii. p. 647 (1822).

Ptilophapa chrysocephala, Lewin, Birds of New Holland, p. 10, pl. 6 (1808).

Turdus melinus, Steph. in Shaw's Gen. Zool. x. p. 262 (1817).

Golden-crowned Honey-eater, Lath. Gen. Hist. iv. p. 194 (1822).

Oriolus regens, Quoy et Gaim. Voy. Uranie, p. 105, pl. xvii. (1824).—Temm. Pl. Col. ii. pl. 320 (1823).—
Wagl. Syst. Av., Oriolus, sp. 7 (1827).

Sericulus chrysocephalus, Swains. Zool. Journ. i. p. 476 (1825).—Vig. & Horsf. Trans. Linn. Soc. xv. p. 328
(1826).—Steph. in Shaw's Gen. Zool. xiv. p. 266 (1826).—Jard. & Selby, Ill. Orn. i. pls. xviii., xix., xx.
(1827).—Less. Traité d'Orn. p. 310 (1831).—Swains. Classif. B. ii. p. 237 (1837).—Gould, B. Austr.
iv. pl. 12 (1843).

Sericulus regens, Less. Voy. Coquille. Zool. i. p. 640, pl. 20 (1826).—Id. Man. Orn. i. p. 256 (1828).—Id. Ois.
Parad., Syn. p. 29.—Id. Hist. Ois. Parad. pls. 26, 27 (1835).

Sericulus auguricollis, Gould, P. Z. S. 1837, p. 145.

Sericulus melinus, Gray, Gen. B. i. p. 233 (1845).—Bp. Consp. Av. i. p. 349 (1850).

Sericulus melinus, Gould, Handb. B. Austr. i. p. 456 (1865).—Ramsay, Ibis, 1866, pp. 325, 330.—Scld. Mus.
Paye-Bas, Coraces, p. 99 (1867).—Ramsay, Ibis, 1867, pp. 415, 456.—Gray, Handb. B. i. p. 293, no. 4335
(1869).—Elliot, Monogr. Parad. pl. xxxx. (1873).—Ramsay, Proc. Linn. Soc. N. S. W. ii. p. 188
(1878).—Sharpe, Cat. Birds Brit. Mus. vi. (1881).—Ramsay, Proc. Linn. Soc. N. S. W. (2) i. p. 1138,
pl. xix. fig. 4 (1887).—North, t.c. pp. 1106, 1173.—Ramsay, Tab. List Austr. B. p. 11 (1888).—
North, Proc. Linn. Soc. N. S. W. (2) iii. p. 1760 (1888).—Id. Nests & Eggs Austr. B. p. 184 (1889).—
Id. Proc. Linn. Soc. N. S. W. (2) v. p. 567 (1891).—Campbell, Proc. R. Soc. Victoria, new ser. v.
p. 128 (1893).—Sharpe, Bull. Brit. Orn. Club, iv. p. xiv (1894).

This splendid Bower-bird is one of the best known of the whole family, as it is one of the finest and most conspicuous species. It appears to be strictly confined to Eastern Australia, where Dr. Ramsay gives its range as New South Wales, and the Wide Bay, Richmond, and Clarence river districts. Mr. Gould's experience of the distribution of the species was as follows:—"It is occasionally seen in the neighbourhood of Sydney, which appears to be the extent of its range to the southward and westward. I met with it in the brushes at Maitland in company with, and feeding on the same trees as, the Satin and Cat-Birds, as well as the Green Oriole (*Mimeta viridis*). It is still more abundant on the Manning, at Port Macquarrie, and at Moreton Bay. I sought for and made every inquiry for it at Illawarra, but did not meet with it, and was informed that it was never seen there ; yet the district is precisely similar in character to those in which it is abundant, about two degrees to the eastward. While encamped on Mosquito Island, near the mouth of the River Hunter, I shot several, and observed it to be numerous on the neighbouring islands, particularly Fisher's Island, where there is a fine garden, and where it commits serious injury to the fruit-crops." Mr. A. J. Campbell has recorded a specimen from Drawings, near the Fitzroy River, and he believes this to be the most northern range of the species.

Everywhere it appears to be the case that the males in full plumage are much more shy and difficult to procure, while the females and young males are not only much rarer, but are very much more plentiful. Mr. Gould never succeeded in obtaining a nest of the Regent Bird ; but the nest was found at Moreton Bay by Mr. F. Strange, who described it as " rudely constructed of sticks, no other material being employed, not even a few roots as a lining." Dr. E. P. Ramsay found a nest at Tarango Creek, in the Richmond River district, built in a cluster of " Lawyer Vines" (*Calamus australis*). In shape the nest was like that of *Collyriocincla harmonica*, and composed of twigs, mosses, leaves, &c., about five inches across by three deep. The bower was first described by Mr. C. Coxen, who received the details from Mr. Waller, of Brisbane, a well-known Australian naturalist and collector. While shooting in a scrub on the banks of the Brisbane River, Mr. Waller saw a male Regent Bird " playing on the ground, jumping up and down, puffing out its feathers, and rolling about in a very odd manner, which occasioned much surprise, as he had never seen the bird on the ground before. The spot where it was playing was thickly covered with small shrubs, and not wishing to lose the opportunity of procuring a specimen he fired, but only succeeded in wounding it; and on searching the spot, he found a bower, formed between, and supported by, two small bresh-plants, and surrounded by small shrubs, so much so, that he had to creep on his hands to get to it. While doing so, the female bird came down from a lofty tree, uttered her peculiar note, and sat on a branch immediately over the bower, apparently with the intention of alighting in front of it, but was scared away by seeing Mr. Waller so close to her. She continued flitting over the place and calling for her mate so long as he was in the neighbourhood. . . . . . The ground around the bower was clear of leaves for some twelve or eighteen inches, and had the appearance of having been swept, the only objects in its immediate vicinity being a small species of *Helix*. The structure was alike at both ends, but the part designated as the front was more easy of approach, and had the principal decorations, the approach to the back being more closed by scrub."

Dr. E. P. Ramsay has given the following account of his discovery of a bower of the Regent Bird :—
"On the 2nd of October, 1866, when returning to our camp, some twenty miles from the township, I stopped to look for an *Atrichia*, which three days before I had heard calling at a certain log; and while standing, gun in hand, ready to fire as soon as the bird—which was at that moment in a remarkably mocking humour—should show itself, I was somewhat surprised at seeing a male Regent Bird fly down and sit within a yard of me. Between the two I hardly knew which choice to take—the *Atrichia*, which was singing close in front of me, or the chance of finding the long-wished-for bower. I decided on the former, and remained motionless for full five minutes, while the Regent Bird hopped round me, and finally on to the ground at my feet, when, looking down I saw the bower scarcely a yard from where I was standing ; had I stepped down off the log I must have crushed it. The bird, after hopping about and rearranging some of the shells (*Helices*) and berries, with which its centre was filled, took its departure, much to my relief, for I was beginning to feel uncomfortable with standing so long in the same position. Further research was not very successful; we met with only one other bower. Wishing to obtain a living specimen of so beautiful a bird as the adult male of this species, I determined to leave the structure until the last thing on my final return to Lismore, which was on the 3rd of November following. We then stopped on our way, and setting eight snares round the bower, anxiously awaited the result. It was not long before we heard a harsh scolding cry of the old bird, and knew that he had 'put his foot into it.' Having taken him out and transferred him to a temporary cage, we carefully pushed a board, brought for the purpose, underneath the bower, and removed it without injury. It is now before me, and is placed upon and supported by, a platform of sticks, which, crossing each other in various directions, form a solid foundation, into which the upright twigs are stuck. This platform is about fourteen inches long by ten broad, the upright twigs are some ten or twelve inches high, and the entrances four inches wide. The middle measures four inches across, and is filled with land-shells of five or six species, and several kinds of berries of various colours, blue, red, and black, which gave it when fresh a very pretty appearance. Besides these there were several newly-picked leaves and young shoots of a pinkish tint, the whole showing a decided 'taste for the beautiful' on the part of this species."

The egg was unknown to Mr. Gould, but a specimen taken from the oviduct of a female is thus described by Mr. North :—" The only egg known of this species at present, which was taken from the oviduct, is in the Museum collection, and is of a long oval, slightly swollen at one end, the ground-colour being of a pale lavender; upon the larger end and beneath the surface of the shell is a zone of nearly round and oval-shaped spots of a uniform pale lilac colour, which in some places are confluent ; on the outer surface, all over the larger end to the lower edge of the zone, are irregularly shaped but well-defined linear markings of sienna, assuming strange shapes; two prominent markings being a double loop and a scroll, others less conspicuous are in the shape of the letter Z and the figure 6, while several of the markings stand at right angles to one another ; from the lower edge of the zone, and dispersed over the rest of the surface, are a few bold dashes of the same colour, several lines being straight, but marked obliquely across the egg, others are like the letter V, with one side lengthened at a right angle, and the figure 7, while upon the lower apex is a single mark in the shape of the letter M. The peculiarity of the markings of this egg is that the spots appear to be on the *under* surface, and the linear markings on the *outer* surface of the shell. Length 1·35 inch × 0·9 inch."

I am indebted to Mr. A. J. Campbell, of Victoria, for the following account of the eggs and bower of the present species. He writes :—

"The Regent Bird, especially the adult male with glorious black and yellow plumage, as Gould has well said, is one of the finest species of the Australian fauna. Last November I undertook an excursion to the Richmond River district, N. S. Wales, with a view of obtaining, amongst other items, the eggs of the Regent Bird. I found the luxuriant scrubs abounding with the birds; in fact they were as plentiful there as the Wattle-birds about the Banksia groves of our southern coast. We experienced no difficulty in procuring our few specimen skins. All that was necessary was to select a balmy day and recline under a *Castilis* tree, where the birds, males in various stages of plumage and females, came to regale themselves on the bunches of hard yellow berries. Nevertheless, although well aided by a hardy companion, I prosecuted a vigorous and toilsome search through dense labyrinths of hot scrub and thorny brakes of prodigal growth, where the thick foliage of the trees caused a perpetual twilight underneath, I returned without the eggs. It was an experience akin to seeking for the proverbial needle in a haystack. From evidence gained by dissection and otherwise it appeared that November was too early for the majority of the birds. However, just prior to leaving on the 19th we detected a female carrying a stick, and after much laborious work we succeeded in tracing her through an entanglement of wild raspberries and stinging trees, and were satisfied that she was building in a certain bushy Bauyong (*Tarrietia*) tree, as we saw her return there several times, each time with a twig in her bill. Marking the tree we pointed it out to two young farmers, directing them to send the eggs after us. Some weeks afterwards I received a doleful letter stating they were unable to climb the tree. However, the next month another farmer, whose scrub-paddock I had scoured, followed up my instructions, and found therein a Regent Bird's nest containing a pair of fresh eggs, which I now have pleasure in describing.

"*Specimen A.* A beautifully well-shaped specimen, with a fine texture of shell of a light yellowish stone colour, with a faint greenish tinge, and marked with blotches and spots of sienna, but principally with hair-like markings of the same colour in fanciful shapes and figures, as if a person had painted them on with a fine brush. Intermingled are a few greyish streaks, dull, as if under the shell's surface. All the markings are fairly distributed, but are more abundant around the upper quarter of the egg. The dimensions are 4 cm, long by a breadth of 2·8 cm, somewhat large compared with the size of the parent. The markings much resemble those of the egg of its close ally the Spotted Bower-bird (*Chlamydodera maculata*), which I found near Wentworth, River Darling, October 1887, with the difference that the ground-colour of the Regent is more yellowish and not of the greenish shade of the Bower-bird.

"*Specimen B.* Similar to the other specimen, but markings less pronounced and finer in character, with a greater proportion of the dull greyish hair-like streaks; also a little smaller—length 3·95 cm, by breadth 2·75 cm.

"The nest was discovered during the last week in December, was placed about 15 feet from the ground, and was observed by the bird sitting on it. The structure was of such a loose nature—merely a few twigs forming a flat shelf about five inches across—that it fell to pieces on removal from the tree. It was accounted remarkable how the eggs could retain their position in it. The description of the nest verifies the statement found in Gould that 'it is rudely constructed of sticks, no other material being employed, not even a few roots as a lining,' but is at variance with Mr. North's statement, which precedes his description of the egg taken from the oviduct of a bird by Mr. Cockerell, the collector, the only other egg at present known."

Adult male. Head, neck, and upper mantle of a velvety texture, and of a brilliant orange-yellow, deepening on the crown into reddish orange; remainder of mantle and entire back, as well as the tail, black; wing-coverts black, as also the primary-coverts; first two primaries black, remainder of the primaries orange-yellow, except along the outer webs and at the tip, encroaching down the inner web; secondaries orange-yellow, tipped with black, excepting the innermost; lores, eyebrow, sides of face and neck, as well as the entire under surface of the body and under tail-coverts, black; under wing-coverts black, the greater series golden yellow, like the lining of the quills; "bill yellow, legs and feet black; iris pale yellow" (J. Gould). Total length 9·5 inches, culmen 1·15, wing 5·1, tail 4·75, tarsus 1·45.

Adult female. Different from the male. General colour above brown, mottled with white centres to the feathers, edged with black; scapulars like the back; wing-coverts and quills plain brown, the latter dusky brown on the inner webs, the innermost secondaries with an irregular white spot at the tip; upper tail-coverts brown, the longer ones with a mesial streak of whitish; tail-feathers brown, more dusky on the inner web; forehead light brown, mottled with minute dusky tips to the feathers; hinder crown and occiput black; sides of head, eyebrow, and nape reddish brown, mottled with dusky edges to the feathers;

hind-neck whitish with dusky margins, followed by a patch of black across the lower hind-neck; lores and base of forehead buffy whitish; cheeks reddish brown, like the sides of the face; chin and sides of throat light reddish, with the centre and lower part of throat black; remainder of under surface of body whity brown, uniform on the abdomen; the breast and sides of the body spotted with blackish-brown margins to the feathers; the thighs and under tail-coverts reddish brown; under wing-coverts and axillaries like the breast, and barred across with dusky brown; quills brown below, light reddish along the inner web: "bill and feet black; iris brown" (*J. Gould*). Total length 10·5 inches, culmen 1·15, wing 5·35, tail 4·5, tarsus 1·5.

The figures in the Plate represent an adult male and female of the size of life, from specimens in the British Museum. The descriptions are taken from the same birds.

BOWER OF THE REGENT BIRD.

From a photograph sent by A. J. Campbell, Esq.

1. CHLAMYDODERA CERVINIVENTRIS. *Gould*
2.   "   MACULATA *(Gould)*

CHLAMYDODERA CERVINIVENTRIS, *Gould.*

Fawn-breasted Bower-bird.

Chlamydera cerviniventris, Gould, P. Z. S. 1850, p. 201.—Macgill. Voy. Rattlesnake, ii. p. 357 (1852).—Gray,
P. Z. S. 1858, p. 194.—Id. Cat. B. New Guinea, p. 59 (1859).—Gould, B. Austr. Suppl. pl. 36 (1859).—
Gray, Hand-l. B. i. p. 394, no. 4342 (1869).—Diggles, Orn. Austr. i. p. 52, pl. 33 (1877).

Chlamydodera cerviniventris, Gould, Handb. B. Austr. i. p. 454 (1865).—Elliot, Monogr. Parad. pl. xxxvi.
(1873).—Salvad. & D'Albertis, Ann. Mus. Civic. Genov. vii. p. 829 (1875).—Id. op. cit. viii. p. 39
(1875).—Id. op. cit. ix. p. 193 (1876).—Sharpe, Journ. Linn. Soc. xiii. p. 91 (1876).—Ramsay, Proc.
Linn. Soc. N. S. W. i. p. 393 (1876).—Sharpe, Journ. Linn. Soc. xiv. p. 495 (1877).—Id. tom. cit.
p. 687 (1879).—Ramsay, Proc. Linn. Soc. N. S. W. ii. p. 188 (1878).—Id. op. cit. iii. pp. 194, 268
(1879).—D'Albertis, Nuova Guinea, pp. 237, 584 (1880).—Salvad. Orn. Papuasia, etc. ii. p. 604
(1881).—Sharpe, Cat. B. Brit. Mus. vi. p. 393 (1881).—Musschenbr. Dagboek, p. 214 (1883).—Finsch,
Vög. der Südsee, p. 17 (1884).—North, Proc. Linn. Soc. N. S. W. (2) i. p. 1165 (1887).—Ramsay,
Tab. List Austr. B. p. 11 (1888).—North, Descr. Cat. Nests & Eggs B. Austr. p. 190, pl. xi. fig. 4
(1889).—Salvad. Aggiunte Orn. Papuasia, etc. ii. p. 165 (1890).—Id. Ann. Mus. Civic. Genov. (2) ix.
p. 585 (1890).—De Vis, Ann. Queensland Mus. no. 2, p. 9 (1892).—Crowley, Bull. Brit. Orn.
Club. i. p. xvi (1892).—Madarász, Aquila, 1894, p. 92.—Sharpe, Bull. Brit. Orn. Club, iv. p. xiv
(1894).—Salvad. Ann. Mus. Civic. Genov. (2) xvi. p. 113 (1896).—Madarász, Term. Füz. 1897,
p. 28.—Reichenow, J. f. O. 1897, p. 214.

Ptilonorhynchus cerviniventris, Giebel, Thes. Orn. i. p. 547 (1872).

Ptilonorhynchus cerviniventris, Rausch, Mitth. orn. Ver. Wien, 1880, p. 54.

This species, which is easily recognized by its fawn-coloured under surface and by the absence of any nuchal frill or ornament in the male, was discovered in the Cape York Peninsula by the late John Macgillivray, during the voyage of the 'Rattlesnake.' It also occurs in South-eastern New Guinea and appears not to be rare in the Port Moresby district, whence I have seen a number of specimens collected by Goldie, Broadbent, and other well-known naturalists. D'Albertis procured it in Hall Bay and at Naiabui, and Dr. Lorin has sent a large series from Kapa Kapa to the east of Port Moresby. The species likewise occurs in German New Guinea on the Finisterre and Bismarck Mountains. It should be noticed that Gray gives the islands of Torres Straits as another locality for the species, but, like Count Salvadori, I have not been able to find any authority for the statement. Mr. De Vis, however, records a specimen from Sariset Island, in the Louisiade Archipelago, obtained by Sir William MacGregor on the 30th of June, 1894.

Dr. E. P. Ramsay observes that this Bower-bird appears to be one of the most common birds about Port Moresby, but it is confined to the coast and is not met with inland. Mr. Masters obtained a bower among the mangroves on the margin of a scrub within the influence of the spring-tides. Mr. Goldie also obtained bowers during his first expedition; they were made of fine twigs placed in an upright or slightly slanting position, and gently arched over in the middle; the inside and sides of the bower, and sometimes the tops of the twigs, were ornamented with berries. The Fawn-breasted Bower-bird is usually found in small troops of six to ten in number, and feeds on fruits and berries.

Gould, in describing the bower found by Macgillivray at Cape York, says:—" It differs from those of the other species; its walls, which are very thick, being nearly upright, or but little inclining towards each other at the top, so that the passage through is very narrow; it is formed of fine twigs, is placed on a very thick platform of thicker twigs, is nearly 4 feet in length and almost as much in breadth, and has here and there a small snail-shell or berry dropped in as a decoration."

The following is Macgillivray's account of his discovery of the species:—

"Two days before we left Cape York, I was told that some Bower-birds had been seen in a thicket or patch of low scrub, half a mile from the beach; and after a long search I found a recently-constructed bower, 4 feet long and 18 inches high, with some fresh berries lying upon it. The bower was situated near the border of the thicket, the bushes composing which were seldom more than 10 feet high, growing in smooth sandy soil without grass.

"Next morning I was landed before daylight, and proceeded to the place in company with Paida, taking with us a large board on which to carry off the bower as a specimen. I had great difficulty in inducing my friend to accompany me, as he was afraid of a war party of Gomokudina, which tribe had lately given notice that they were coming to fight the Evans Bay people. However, I promised to protect him, and loaded one barrel with ball, which gave him increased confidence; still he insisted upon carrying a large bundle of spears and a throwing-stick.

"While watching in the scrub, I caught several glimpses of the *Tsunga* (its native name) as it darted through the bushes in the neighbourhood of the bower, announcing its presence by an occasional loud *chuer-r-r*, and imitating the notes of various other birds, especially the *Tropidorhynchus*. I never before met with a more wary bird; and, for a long time, it enticed me to follow it to a short distance, then flying off and alighting on the bower it would deposit a berry or two, run through and be off again before I could reach the spot. All this time it was impossible to get a shot. At length, just as my patience was becoming exhausted, I saw the bird enter the bower and disappear, when I fired at random through the twigs, fortunately with effect. So closely had we concealed ourselves latterly, and so silent had we been, that a kangaroo, while feeding, actually hopped up within fifteen yards, unconscious of our presence until I fired at."

Eggs of the present species are in the Australian Museum and in the collection of Mr. Philip Crowley. The latter were taken by Mr. A. Goldie in Milne Bay, S.E. New Guinea.

Mr. A. J. North (Proc. Linn. Soc. N. S. Wales, (2) i. p. 1160) says that the egg is very like that of *C. maculata* in colour, with the same peculiar linear markings crossing and recrossing each other all round, but confined more to the larger end of the egg than is usually the case with *C. maculata*. A specimen in the Australian Museum collection, taken at Cape York, measures 1·4 inch in length by 1·03 inch in breadth. The nest is an open one, cup-shaped, and built near the ground; it is composed of twigs, pieces of bark and moss, and is lined inside with grass &c."

The following description of the species is copied from my sixth volume of the 'Catalogue of Birds':—

Adult. Above brown, all the feathers edged with ashy, giving a greyish shade to the upper parts, nearly uniform on the hind-neck; crown of head, feathers above the eyes, and lores thickly but minutely dotted with triangular spots of buffy white; the whole of the back, scapulars, and wing-coverts distinctly streaked down the shaft with buffy white, dilating into a triangular spot at the tip, all the apical markings much larger and whiter on the wing-coverts, the primary-coverts edged with whitish near the tip; quills brown, externally washed with greyish, the secondaries tipped with white, forming a large spot at the tip of the innermost; rump and upper tail-coverts streaked like the back, but slightly more tinged with fulvous; tail-feathers brown, washed with greyish along the edge of the outer webs and tipped with white; entire sides of face and throat ashy brown, thickly streaked everywhere with light fawn-buff, all the feathers being axially streaked with this colour; chest fawn-buff, mottled with ashy brown, with which colour the feathers are edged and slightly barred; all the rest of the under surface of the body clear fawn-colour, the flanks indistinctly mottled with indications of ashy-brown bars; sides of the upper breast brown, broadly streaked down the centre with fulvous; under wing-coverts fawn, like the underparts, the outermost of the greater series ashy brown, with pale fulvous bases, the lower surface of the quills light brown, edged with pale fulvous along the inner web; "bill black" feet grey; iris black" (*J. Macgillivray*); "feet greenish; iris dark maroon" (*L. Loria*). Total length 11·3 inches, culmen 1·1, wing 5·65, tail 4·91, tarsus 1·7.

The upper figure in the accompanying Plate represents an adult male of this species from Cape York, of the size of life.

CHLAMYDODERA MACULATA (*Gould*).

Spotted Bower-bird.

Calodera maculata, Gould, P. Z. S. 1836, p. 106.—Id. Syn. B. Austr. part i. (1837).

Chlamydera maculata, Gould, B. Austr. part i. (1837, cancelled).—Id. op. cit. iv. pl. 8 (1841).—Gray, Gen. B. ii.
    p. 225 (1846).—Bp. Consp. Av. i. p. 370 (1850).—Diggles, Orn. Austr. i. p. 52, pl. 52 (c. 1867).
    —Gray, Hand-l. B. ii. p. 294, no. 6342 (1869).

Chlamydodera maculata, Cab. Mus. Hein. Th. i. p. 212 (Oct. 1851).—Gould, Handb. B. Austr. i. p. 450 (1865).
    —Ramsay, Ibis, 1866, p. 329.—Elliot, Monogr. Parad. pl. xxx. (1873).—Ramsay, P. Z. S. 1874, p. 605.
    —Id. Proc. Linn. Soc. N. S. W. ii. p. 188 (1878).—Sharpe, Cat. B. Brit. Mus. vi. p. 389 (1881).—
    Ramsay, Proc. Linn. Soc. N. S. W. vii. p. 409, pl. iii. fig. 2 (1883).—North, op. cit. (2) i. pp. 1157,
    1164 (1887).—Ramsay, Tab. List Austr. B. p. 11 (1888).—North, Descr. Cat. Nests and Eggs
    Austr. B. p. 178, pl. xi. fig. 3 (1889).—Sharpe, Bull. Brit. Orn. Club, iv. p. xlv (1894).

Ptilorhynchus maculatus, Schl. Mus. Pays-Bas, Coraces, p. 119 (1867).

THE Spotted Bower-bird is distinguished by the reddish spots or bars at the tips of the feathers of the upper surface, which give the bird a strongly mottled appearance, as well as by the dusky spots and bars on the flanks and throat. The head is rufous brown, varied with blackish edgings and spots on the feathers, and the male has a lilac band on the nape.

Mr. E. P. Ramsay, in his 'Tabular List of Australian Birds,' gives the range of the species as from Cape York and Rockingham Bay to Port Denison, the Dawson River, and the Wide Bay district, as well as New South Wales, the interior of Australia to Victoria, and South Australia. Mr. A. G. North also adds the Clarence River district as a habitat of the species, so that its range is complete from Cape York to New South Wales and thence west to Victoria and South Australia.

In his 'Descriptive Catalogue of the Nests and Eggs of Birds found breeding in Australia and Tasmania,' Mr. North writes:—" Our knowledge of the range of this species has recently been extended to Cape York. Previously Rockingham Bay was considered its northern limit on the coast, and the Murray district is Victoria and South Australia its most southern range. The interior provinces are the stronghold of this species, where it is found plentifully dispersed all over the Lachlan and Darling River districts. It also occurs inland eighty miles west from Rockhampton."

Gould has given the following account of the species in his 'Handbook to the Birds of Australia' :—

" During my journey into the interior of New South Wales, I observed this bird to be tolerably abundant at Brezi on the river Mokai to the northward of the Liverpool Plains; it is also equally numerous in all the low scrubby ranges in the neighbourhood of the Namoi, as well as in the open brushes which intersect the plains on its borders; and collections from Moreton Bay generally contain examples, still from the extreme shyness of its disposition, the bird is seldom seen by ordinary travellers, and it must be under very peculiar circumstances that it can be approached sufficiently close to observe its colours. The Spotted Bower-bird has a harsh, grating, scolding note, which is generally uttered when its haunts are intruded on, and by this means its presence is detected when it would otherwise escape observation. When disturbed it takes to the topmost branches of the loftiest trees, and frequently flies off to another neighbourhood.

" In many of its actions and in the greater part of its economy much similarity exists between this species and the Satin Bower-bird, particularly in the curious habit of constructing an artificial bower or playing-place. I was so far fortunate as to discover several of these bowers during my journey in the interior, the finest of which I succeeded in bringing to England; it is now in the British Museum. The situations of these runs or bowers are much varied: I found them both on the plains studded with Myalls (*Acacia pendula*) and other small trees, and in the brushes clothing the lower hills. They are considerably longer and more avenue-like than those of the Satin Bower-bird, being in many instances three feet in length. They are outwardly built of twigs, and beautifully lined with tall grasses, so disposed that their heads nearly meet; the decorations are very profuse, and consist of bivalve shells, crania of small mammalia and other bones bleached by exposure to the rays of the sun or from the camp-fires of the natives. Evident indications of high instinct are manifest throughout the bower and decorations formed by this species, particularly in the manner in which the stones are placed within the bower, apparently to keep the grasses with which it is lined fixed firmly in their places; these stones diverge from the mouth of the run on each side so as to form little paths, while the immense collection of decorative materials are placed in a heap before the entrance of the avenue, the arrangement being the same at both ends. In some of the larger bowers, which had evidently been resorted to for many years, I have seen half a bushel of bones, shells, &c., at each of the entrances. I frequently found these structures at a considerable distance from the rivers, from the borders of which they could alone have procured the shells and small round pebbly stones; their collection and transportation must therefore be a task of great labour. I fully ascertained that these runs, like those of the Satin Bower-bird, formed the rendezvous of many individuals; for, after secreting myself for a short space of time near one of them, I killed two males which I had previously seen running through the avenue."

Mr. North tells me (in epist.) that " this Bower-bird thrives well in confinement, and its powers of mimicry rival those of *Menura*."

Dr. Ramsay writes (Proc. Linn. Soc. N. S. W. vii. p. 409) :—

" I have received this species of Bower-bird from almost every part of the interior of Queensland, New South Wales, and South Australia, and eggs from the Dawson River in Queensland, the Barkoo, the Clarence River, and from the Cobar district in New South Wales. They differ very little in the tints of the markings, varying in shades of umber, sienna, and olive-brown. Those at present under consideration were taken by Mr. James Ramsay in the Cobar district; they are of a pale greenish white with numerous thick lines of umber wound round the whole surface, irregular, wavy, crossing and recrossing here and there, forming loops and knots, and occasionally crossed by a line of black or an obsolete line of olive or slaty brown. The nest is an open structure of sticks and grasses, round, about five inches inside diameter, by three deep, and four inches high; it is placed between the thick upright forks of a tree. The eggs are two to three in number for a sitting, length 1·53 inch × 1·07 inch in diameter."

Mr. North, in his 'Catalogue,' observes :—" The nest is an open structure, usually placed in a low tree, and is saucer- or bowl-shaped, composed of sticks and lined with grass, about five inches inside diameter by three inches deep, and four inches high. It is very rarely indeed that C. maculata is found near the coast, although on one occasion Dr. Ramsay procured an egg on Ash Island, near Hexham, on the Hunter River, about ten miles from the sea-coast. This was in 1861, and probably the first time that the egg had been found, though this fact appears to have escaped Dr. Ramsay's memory, since he described another egg of the same species thirteen years afterwards (P. Z. S. 1874, p. 605), when Mr. J. B. White was credited with having obtained the first specimen.

" In 1875 Mr. James Ramsay procured several specimens of both birds and eggs near Tyndarie; and others were received from the Clarence River district. Since then the eggs have become less rare in collections, and are to be found in most of those formed in the interior. The eggs of C. maculata vary considerably in the extent of their markings, and sometimes in the tints of colouring. One I have from the Dawson River district is slightly smaller than usual, and has the ground-colour of a faint greenish-grey, covered all over with a fine network of light brownish linear markings, closer together near the thickest end; others have their markings confined altogether to the larger end of the egg. A set taken by Mr. John Macgillivray at Grafton on the Clarence River, on the 7th of September, 1864, measures :—Length 1·47–1·5 inch × 1·09 inch."

Mr. North has kindly sent me a photograph of two eggs taken by Mr. James Ramsay at Tyndarie, and describes them as follows :—" They are of a greenish-grey colour, which is almost obscured by numerous linear markings, short streaks, and fine hair-lines of umber-brown. One specimen has two black linear streaks on the larger end and a few indistinct clouded blotches of pale violet-grey appearing as if beneath the surface of the shell. Length 1·47–1·5 inch × 1·09 inch."

In a letter recently received from Mr. North he writes :—" A correspondent of mine, whose accuracy in these matters I can vouch for, informed me that he saw a most remarkable bower of C. maculata near Cobar in N. S. Wales. It was formed of curved twigs as usual, which met near the top and, recurving again, formed a second bower above, much smaller than the one underneath. The lower bower measured about two feet, the one on the top (which was in the centre) one foot. It had the usual complement of bones, also a few of Eley's cartridge-cases." Mr. North sent me a rough sketch of the bower, from which Mr. H. Grönvold has drawn a little picture (see opposite page).

The following descriptions are taken from my sixth volume of the 'Catalogue of Birds' :—

Adult male. General colour above dark brown, spotted all over with tawny buff near the end of each feather, these spots paling into whitish near the apex; the hind-neck plain umber-brown, separating the nape-spot from the mantle; head tawny buff, mottled and, as it were, striped with dark brown edges to the feathers; a few of the feathers of the crown with silvery-whitish tips; on the nape a band of beautiful lilac plumes, somewhat elongated laterally; wing-coverts like the back, and spotted with tawny buff at the tips; the quills brown, edged with whity brown and having pale spots at the end of the secondaries, these spots being somewhat obsolete on the tips of the primaries; upper tail-coverts blackish brown, having in addition to the fulvous bar at the end, a second sub-terminal bar of tawny buff; tail-feathers brown, edged with pale brown along both webs, and tipped with pale tawny buff; lores and ear-coverts, as well as the sides of the neck, tawny buff, like the head, and striped in the same manner, with dusky brown edges to the feathers; cheeks buffy whitish, the feathers edged with brown; throat fulvescent, shading off into whity brown on the chest, the breast and abdomen being uniform creamy buff; the sides of the body whitish, barred with dusky on the flanks and thighs; the throat and fore-neck spotted with small bars of dusky; under tail-coverts pale tawny buff, with a few remains of dusky brown bars; axillaries buffy whitish; under wing-coverts tawny buff, with a few dusky brown bars; quills ashy brown below, pale yellow along the inner web: "bill and feet dusky brown; bare skin at the corner of the mouth thick, fleshy, prominent, and of a pinky flesh-colour; iris dark brown" (*Gould*). Total length 11·5 inches, culmen 1·1, wing 5·75, tail 4·25, tarsus 1·6.

Adult female. Very similar to the male, but without the lilac-coloured band on the nape, the latter being of the same colour as the head, and separated from the mantle by the hind-neck, which is of a lighter and more umber-brown than the back. Total length 11·5 inches, culmen 1·1, wing 5·7, tail 4·5, tarsus 1·6.

The figure in the Plate represents a male of this species drawn from a specimen in the British Museum.

BOWER OF THE SPOTTED BOWER-BIRD.

From a sketch by Mr. Stuart.

CHLAMYDODERA OCCIPITALIS, Gould.

Large-frilled Bower-bird.

Chlamydodera occipitalis, Gould, Annals & Mag. Nat. Hist. 4th series, xvi. p. 429 (1875).—Id. B. New Guin. i. pl. 45 (1879).—Ramsay, Proc. Linn. Soc. N. S. W. ii. p. 188 (1878).

Chlamydodera maculata (pt.), Sharpe, Cat. Birds in Brit. Mus. vi. p. 389 (1881).—Ramsay, Tab. List Austr. B. p. 11 (1888).

THIS species was described by the late Mr. Gould in 1875, from a specimen procured at Port Albany, Northern Australia. He separated it from *C. maculata* on account of its "somewhat larger size and the extreme beauty of its occipital patch, which is nearly twice as large as in the species mentioned, and is even of a more brilliant lilac-colour, particularly if the frill be turned up and seen from beneath."

Dr. E. P. Ramsay examined the type of this species when he was last in England, and came to the conclusion that it was nothing more than a very fine male specimen of the ordinary *C. maculata*, and when I wrote the sixth volume of the 'Catalogue of Birds' I came to the same conclusion, after an examination of the type, which had by that time become the property of the British Museum. As, however, the bird is certainly much finer than any other specimen I have seen, and as its habitat is considerably removed from that of *C. maculata*, I have considered it worth refiguring in the present Monograph under the name bestowed upon it by Mr. Gould, trusting to the zeal of Australian naturalists to discover more specimens and decide the differences, if there are any, between the two Spotted Bower-birds. Mr. Gould never accepted my opinion about his *C. occipitalis*, and affirmed its specific distinctness up to the day of his death. I think, therefore, that it would have been wrong to have omitted a figure of the bird in the present work.

The Plate represents the adult male, and is a reproduction of the one published by Mr. Gould in the 'Birds of New Guinea.' He adds:—" The decorative bower forming part of the illustration is taken from a photograph of some unknown species sent to me by the late Mr. Cozen, of Brisbane; it may or may not be that of the present bird."

CHLAMYDODERA NUCHALIS (*Jard. & Selby*).

Western Lilac-naped Bower-bird.

Ptilonorhynchus nuchalis, Jard. & Selby, Illustr. Orn. ii. pl. 102 (1826).

Calodera nuchalis, Gould, Syn. B. Austr. pt. i. (1837).

Chlamydera nuchalis, Gould, B. Austr. pt. i. (cancelled).—Id. op. cit. iv. pl. ix. (1841).—Stokes, Discov. Austr. ii. p. 97 (1846).—Gray, Gen. B. ii. p. 235 (1846).—Bonap. Consp. Av. i. p. 370 (1850).—Jacq. et Pucher. Voy. Pôle Sud, texte, iii. p. 64 (1853).—Gray, Hand-l. B. i. p. 294, no. 6339 (1869).

Chlamydère à nuque orneé, Hombr. et Jacq. Voy. Pôle Sud, Atlas, pl. vii. fig. 2 (1842-53).

Chlamydodera nuchalis, Cab. Mus. Hein. Th. i. p. 212 (1851).—Gould, Handb. B. Austr. i. p. 448 (1865).—Elliot, Monogr. Parad. pl. xxxi. (1873).—Masters, Proc. Linn. Soc. N. S. W. ii. p. 273 (1878).—Sharpe, Cat. B. Brit. Mus. vi. p. 391 (1881).—Ramsay, Proc. Linn. Soc. N. S. W. (2) ii. p. 169 (1888).—Id. Tab. List Austr. B. p. 11 (1888).—Sharpe, Bull. Brit. Orn. Club, iv. p. xlv (1894).

Ptilochynchus nuchalis, Schl. Mus. Pays-Bas, Coraces, p. 119 (1867).

This species and its eastern ally are distinguished from *Chlamydodera maculata* and its allies by their more uniform upper surface, the feathers being margined with ashy-whitish or having white tips or bars at the ends, the throat and sides of the body being perfectly uniform.

It was first described by Jardine and Selby, and Dr. Ramsay (Tab. List Austr. B. p. 10) believes that the original specimen must have been found " in North-western Australia, probably during Leichardt's Expedition, probably by Gilbert or Elsey, near Port Essington." In the same book Dr. Ramsay gives the distribution of *C. nuchalis* as the Derby district in N.W. Australia, Port Darwin and Port Essington, and the Gulf of Carpentaria.

Gould writes in 1865, in his 'Handbook,' :—This fine species was first described and figured in the 'Illustrations of Ornithology,' by Sir William Jardine and Mr. Selby, from the then unique specimens in the collection of the Linnean Society ; but neither the part of Australia of which it is a native nor any particulars relative to its habits were known to those gentlemen. It is now clearly ascertained that it is an inhabitant of the north-west coast, a portion of the Australian continent that has, as yet, been little visited. I am indebted for individuals of both sexes to two of the officers of H.M.S. 'Beagle,' Messrs. Byrne and Dring ; but neither of these gentlemen furnished me with any account of its economy.

"The following passage from Captain Stokes's 'Discoveries in Australia' (vol. ii. p. 97) comprises all that has been recorded respecting the curious bower constructed by this bird :—

"'I found matter for conjecture in noticing a number of twigs with their ends stuck in the ground, which was strewed over with shells, and their tops brought together so as to form a small bower ; this was 2½ feet long, 1½ foot wide at either end. It was not until my next visit to Port Essington that I thought this anything but some Australian mother's toy to amuse her child ; upon being asked, one day, to go and see the 'birds' playhouse,' I immediately recognised the same kind of construction I had seen at the Victoria River, and found the bird amusing itself by flying backwards and forwards, taking a shell alternately from each side, and carrying it through the archway in its mouth.'"

The following descriptions are copied from my sixth volume of the 'Catalogue of Birds' :—

Adult male. General colour above ashy brown, the feathers being dark brown edged with ashy brown, these margins being very distinct and broad on the rump and upper tail-coverts, the latter having indications of a subterminal whitish spot as well as the light tip ; wing-coverts dark brown, margined with ashy, and tipped with a whitish spot, which is much broader on the secondaries ; tail-feathers brown, margined with ashy on both webs, and barred with whitish at the tips ; head brown, each feather having a minute spot of ashy at the tip ; on the nape a band of beautiful lilac feathers, fringed with some stiffened brown feathers, tipped with silvery whitish ; hind neck uniform ashy brown, separating the nuchal band from the mantle ; lores, sides of face, and under surface of body light ashy brown, becoming paler on the abdomen, where it inclines to creamy white, the lower flanks and thighs having faint indications of dusky bars, these becoming zigzag and distinct in character on the under tail-coverts, which are whitish like the abdomen; under wing-coverts and axillaries pale ashy brown, with a few indications of dusky bars; quills ashy brown below, pale yellowish along the inner web: " bill, legs, and irides brownish " (*Gould*). Total length 14·5 inches, culmen 1·4, wing 6·75, tail 5·5, tarsus 1·9.

Adult female. Differs from the male in wanting the lilac band on the nape, and in having the under surface faintly barred with dusky. Total length 13 inches, culmen 1·4, wing 6·75, tail 5·5, tarsus 1·9.

The figures in the Plate are drawn from specimens in the British Museum.

CHLAMYDODERA ORIENTALIS, *Gould.*

Queensland Lilac-naped Bower-bird.

Chlamydera nuchalis (nec Jard. & Selby), Ramsay, Ibis, 1865, p. 85.

Chlamydodera nuchalis (nec Jard. & Selby), Ramsay, Ibis, 1866, p. 329.—Id. Proc. Zool. Soc. 1878, p. 385.—Id. Proc. Linn. Soc. N. S. Wales, ii. p. 188 (1878).

Chlamydodera orientalis, Gould, Ann. & Mag. Nat. Hist. (5) xr. p. 74 (1879).—Id. Birds of New Guinea, i. pl. 44 (1880).—Sharpe, Cat. B. Brit. Mus. vi. p. 392 (1881).—Ramsay, Tab. List Austr. B. p. 11 (1888).

This is the eastern representative of the Lilac-naped Bower-bird of North-western Australia. It is found only in Queensland and the districts of Port Denison and Rockingham Bay. The differences between the present species and the true *Chlamydodera nuchalis* were first pointed out by Mr. Gould in 1879, and have since been admitted by myself in the 'Catalogue of Birds' and by Dr. E. P. Ramsay in his 'Tabular List of Australian Birds,' where will be found a note on the differences between the eastern and western forms.

Dr. Ramsay states that he has received several specimens from Port Denison, where it is by no means rare. His correspondent, Mr. Rainbird, sent him a living example, which he had in confinement for five months. Dr. Ramsay says :—" It fed freely on bread soaked in water, and on almost anything in the shape of fruit. It was a great mimic, and imitated many of our native birds with much precision, accompanied by the most varied and pleasing actions."

Adult male. Similar to *C. nuchalis*, but much more mottled on the upper surface, with whitish tips to the feathers, these markings being very apparent on the head, which is not so uniform as in *C. nuchalis*. Total length 13·5 inches, culmen 1·5, wing 7·0, tail 5·5, tarsus 1·9.

Adult female. Similar to the male, but wanting the lilac nuchal spot ; the whitish spots on the back also smaller and less pronounced. Total length 12·5 inches, culmen 1·35, wing 6·7, tail 5·2, tarsus 1·7.

Young male. Similar to the adult female at first, and acquiring the nuchal patch by a moult.

The descriptions and figures are taken from the typical specimens in the British Museum.

CHLAMYDODERA GUTTATA, *Gould.*

Large-spotted Bower-bird.

Chlamydera guttata, Gould, P. Z. S. 1862, p. 162.—Id. B. Austr. Suppl. pl. 35 (1857).—Gray, Hand-l. B. i. p. 294, no. 4340 (1869).

Chlamydodera guttata, Gould, Handb. B. Austr. i. p. 452 (1865).—Elliot, Monogr. Parad. Intr. p. 22 (1873).— Ramsay, Proc. Linn. Soc. N. S. W. ii. p. 183 (1878).—Sharpe, Cat. B. Brit. Mus. vi. p. 390 (1881).— North, Proc. Linn. Soc. N. S. W. (2) i. p. 1159 (1887).—Ramsay, Tab. List Austr. B. p. 11 (1888).— Stirling & Zietz, Trans. R. Soc. S. Austr. xvi. p. 157 (1893).—Sharpe, Bull. Brit. Orn. Club, iv. p. xlv (1894).—North, Rep. Horn Sci. Exped. Centr. Austr. part ii. Zool. Aves, p. 90 (1896).

This species differs from *C. maculata* in having the crown of the head silvery brown, slightly tinged with rufous bars, the bases to the feathers being black. Mr. North, describing the two males brought back by the Horn Expedition to Central Australia, says:—"Two examples of this distinct and well-marked species were obtained at Glen Edith. Both are males and apparently not quite adult, or in the moult, for one has only a faint indication of the beautiful blue nuchal plumes, and they are but slightly more developed in the other specimen. This species is readily distinguished from its near ally, *C. maculata* of Eastern and Southern Australia, by the feathers of the upper surface being blackish brown instead of dark brown— rendering the spots, which are paler, more conspicuous—and by the absence of the earthy-brown band between the nuchal plumes and the mantle. The head and neck, too, are much darker, and the tips of the wing-coverts and secondaries are pale yellowish buff, instead of tawny buff."

The following is the note on the species made by Mr. Keartland, the naturalist to the Horn Expedition:— " Wherever the 'native fig' trees existed, these birds were found. They were generally very shy, and only two specimens were obtained. Several bowers seen bore a close resemblance to those of *C. maculata*. At Owen Springs we were informed that in dry weather these birds come to the water-buckets under the veranda to drink, and become quite fearless of the presence of persons sitting close by."

The species was first met with by Mr. Stuart during his journey across the Australian continent from Adelaide to the Victoria River, and the head of an adult male obtained by him is in the Gould collection in the British Museum, in which institution is a perfect skin of a female collected somewhere in North-western Australia.

It is doubtless this species, as Gould has pointed out, which is referred to in his 'Travels' (vol. i. pp. 196, 245) by Sir George Grey, who met with it at the summit of one of the sandstone-ranges forming the watershed of the streams flowing into the Glenelg and Prince Regent's rivers. He writes:—" We fell in with a very remarkable nest, or what appeared to me to be such. We had previously seen several of them, and they had always afforded us food for conjecture as to the agent and purpose of such singular structures. This very curious sort of nest, which was frequently found by myself and other individuals of the party, not only along the sea-shore, but in some instances at a distance of six or seven miles from it, I once conceived must have belonged to a kangaroo, until I was informed that it was the run or playing-place of a species of *Chlamydodera*. These structures were formed of dead grass and parts of bushes, sunk a slight depth into two parallel furrows of sandy soil, and then nicely arched above. But the most remarkable fact connected with them was that they were always full of broken sea-shells, large heaps of which protruded from each extremity. In one instance, in a bower the most remote from the sea that we discovered, one of the men of the party found and brought to me the stone of some fruit which had evidently been rolled in the sea; these stones he found lying in a heap in the nest, and they are now in my possession."

The following is the description of the species given by me in my sixth volume of the ' Catalogue of Birds ' :—

" Very similar to *C. maculata*, but altogether darker above, and having the neck of the same dark brown as the back, with smaller tawny buff spots ; the under surface of the body is also darker. Total length 12 inches, culmen 1·05, wing 5·6, tail 4·25, tarsus 1·6.

"The head of the male, collected by Mr. Stuart during his travels into the interior of Australia, likewise points to the species being distinct from *C. maculata*. The blue band is much richer in tint, and the head shows the whole of the feathers with silvery tips, instead of only a few thus marked as in *C. maculata*. The feathers of the hind-neck resemble those of the female, and seem to indicate that the species has no band of earthy brown between the nape and the mantle as in its near ally."

The species so closely resembles *C. maculata* that a separate figure has been considered unnecessary.

CHLAMYDODERA RECONDITA, *Meyer.*

Meyer's Bower-bird.

Chlamydodera recondita, Meyer Abhandl. k. zool. Mus. Dresden, 1894–95, no. 10, p. 2 (1895).

Dr. A. B. Meyer has described and figured the egg of a species of *Chlamydodera* from Constantine Harbour, in Kaiser Wilhelm's Land in German New Guinea, where it was taken by Herr A. Grubauer.

CHLAMYDODERA LAUTERBACHI, *Reichen.*

Lauterbach's Bower-bird.

Chlamydodera lauterbachi, Reichenow, Orn. Monatsb. v. p. 21 (1897).—Id. J. f. O. 1897, p. 215, pl. vi.

This fine species, which I have not seen myself, but which Mr. Hartert informs me is a true *Chlamydodera*, was discovered by Lauterbach on the Jagei River, a tributary of the Ramu River in German New Guinea. It has been figured by Dr. Reichenow (*l. c.*), and is evidently quite distinct from all the other species of the genus. Whether the egg on which Dr. Meyer founded his *C. recondita* will ultimately be discovered to belong to *C. lauterbachi* is a problem with which we need not trouble ourselves at present.

The following is a translation of Dr. Reichenow's original description :—

Adult male. Crown of head and cheeks golden-orange ; nape yellowish olive-brown ; feathers of the upper part of the body, upper tail-coverts, and lesser wing-coverts olive-brown, with yellowish edgings at the tip ; median and greater coverts olive-brown with a whitish fringe at their tips ; foreneck pale yellow, striated with brown, each feather having a pale yellow shaft-stripe and brown margins ; centre of the throat nearly uniform pale yellow ; under surface of body and under tail-coverts pale chrome-yellow, the flanks banded across with pale brown ; under wing-coverts pale yellow, the longer ones with pale brownish tips ; tail-feathers dark olive-brown, with yellowish outer margins and broad whiter inner margins and tips ; quills dark brown, with narrow pale yellowish outer margins ; the shafts yellow below ; the secondaries with whitish tips : bill black ; foot grey ; iris brown. Length 285 millim., wing 130, tail 110, bill 22, tarsus 40.

There is no specimen of *C. lauterbachi* in England, the type being unique in the Berlin Museum, so I have been unable to give a figure of the species.

ÆLURŒDUS MELANOTIS (Gray).

Black-cheeked Cat-bird.

Ptilonorhynchus melanotis, Gray, Proc. Zool. Soc. 1858, pp. 181, 194.—Id. Cat. B. New Guin. pp. 37, 59 (1859).—Id. P. Z. S. 1861. p. 434. Rosenb. Nat. Tijdschr. Nederl. Ind. xxv. p 236 (1863, pt.).—Id. J. f. O. 1864, p. 122 (pt.). Finsch, Neu-Guinea, p. 173 (1865).—Rosenb. Reis naar Zuidoosternd. p. 47 (1867).—Gray, Hand-l. B. i. p. 294, no. 4338 (1869).—Pelz. Verh. k.-k. zool.-bot. Gesellsch. 1872, p. 428.—Musschenbr. Dagboek, pp. 212, 241 (1883).—Rosenb. Malih. orn. Ver. Wien, 1885. p. 54.

Ptilorhynchus melanotis, Schl. Mus. Pays-Bas. Coracias, p. 116 (1867).—Id. Nederl. Tijdschr. Dierk. iv. p. 54 (1873, pt.).

Ælurœdus melanotis, Meyer, Sitzb. k. Akad. d. Wissensch. lxix. pp. 82, 83 (1874).—Gould, B. New Guinea, i pl. 39 (1875.)

Ælurœdus melanotis, Elliot, Monogr. Parad. pl. 35 (1873).—Id. Introd. Monogr. Parad. p. xxii (1873, pt.).—Scl. Ibis, 1874, p. 416.—Salvad. Ann. Mus. Gen. iv. p. 193 (1874.)—Id. Proc. Zool. Soc. 1876, p. 99.— D'Alb. & Salvad. Ann. Mus. Gen. xiv p. 114 (1879).—D'Alb. Nuova Guin. p. 588 (1880).—Salvad. Orn. Papuasia, ii. p. 673 (1881).—Id. Voy. 'Challenger,' Birds, p. 82 (1881).—Sharpe, Cat. B. Brit. Mus. vi. p. 380 (1881).—Meyer, Zeitschr. ges. Orn. i. p. 293 (1884).—Salvad. Agg. Orn. Papuasia, ii. p. 166 (1890.).

This is the largest of the Cat-birds with white tips to the tail. It was discovered in the Aru Islands by Dr. Wallace during his expedition to the Malay Archipelago, and has since been discovered by Baron von Rosenberg, Mr. Hoedt, and Dr. Beccari in most of the islands of the Aru group—such as Wokan, Trangan, and Maikor ; D'Albertis also collected several specimens on the Fly River, in May and June. It is therefore one of the many species in Southern New Guinea which prove the relationship between the avifauna of the southern portion of the great Papuan Island and that of the Aru Archipelago.

Of the habits of this Cat-bird we know nothing. D'Albertis states that it feeds on fruit, and he found in the stomach the same seeds as those devoured by *Paradisea raggiana*, *Ptilorhis intercedens*, and *Casuarius regius*.

Count Salvadori states that the females are a little smaller than the males and have the ochreous colour of the lower parts less decided, while the margins of the feathers of the upper part of the breast are duller and show less of a green shade. The reddish colour on the feathers of the head and neck, which is sometimes met with, appears to Salvadori to be a sign of youth. D'Albertis describes the colour of the soft parts as follows :—" Bill ruby whitish ; feet ashy ; iris chestnut or coppery red."

Adult male (type of species). General colour above bright grass-green, slightly shaded with blue on the wing-coverts and primaries ; the median and greater wing-coverts and the secondaries tipped with buffy white ; quills dull brown on the inner webs ; tail dull grass-green, blackish on the inner webs of the outer feathers, all the feathers rather broadly tipped with white ; head and neck fulvous, mottled with black, the feathers being mostly of the latter colour, with a large oval spot of fulvous near the tip, very much larger on the hinder neck ; lores and a line of feathers above and below the eye buffy whitish ; behind the eye a bare patch ; ear-coverts black ; rest of the sides of the face buffy white, barred with narrow crescentic cross lines of black ; the fore-neck and chest yellowish white, all the feathers dark at base and distinctly edged with black, giving a very strongly mottled appearance ; rest of under surface of body yellowish buff, slightly inclining to fawn-buff, the feathers with narrow whitish shaft-lines, the breast mottled with sub-terminal cross lines of blackish ; sides of body washed with green, with rather broad mesial streaks of white on the flank-feathers ; under wing-coverts yellowish buff, tinged with green along the edge of the wing, many of the abdominal plumes, when lifted, exhibiting a bluish shade underneath, this being also seen on the lower surface of the tail. Total length 12·5 inches, culmen 1·6, wing 6·65, tail 5·1, tarsus 1·85.

Adult female. Similar to the male, but more distinctly greenish underneath ; each feather with a terminal spot of brighter green, the lower flanks washed with bluish green ; spots on secondaries very large and distinct ; mantle varied with arrow-shaped central markings of yellowish buff to all the feathers. Total length 13·3 inches, culmen 1·45, wing 6·1, tail 5·2, tarsus 1·8.

The figures in the Plate are taken from specimens in the Gould collection, *now* in the British Museum.

ÆLURŒDUS ARFAKIANUS, Meyer.

Arfak Mountain Cat-bird.

? *Ptilonorhynchus melanotis* (nec Gray), Rosenb. Nat. Tijdschr. Nederl. Ind. xxv. p. 236 (1863, pt. : teste Salvadori).—Id. J. f. O. 1864. p. 122 (pt.).

Ptilorhynchus melanotis, pt. (nec Gray), Schlegel, Nederl. Tijdschr. Dierk. iv. p. 53 (1871).

Ailuroedus melanotis (nec Gray), Sclater, Proc. Zool. Soc. 1873, p. 697.—Elliot, Monogr. Parad., Intr. p. xxii (1873, pt.).—Sclater, Ibis, 1874, p. 416.

Ailuroedus arfakianus, Meyer, Sitz. k. Akad. Wissensch. Wien, lxiv. p. 83 (1874).—Sclater, Ibis, 1874, p. 416.—Gould, B. New Guinea, i. pl. 40 (1875).—Salvad. Ann. Mus. Civic. Genov. ix. p. 193 (1876), x. p. 151 (1877).—Id. Proc. Zool. Soc. 1878, p. 99.—D'Albert. Nuova Guinea, p. 581 (1880).

Ptilorhynchus arfakianus, Giebel, Thes. Orn. iii. p. 330 (1877).

Ptilonorhynchus arfakianus, Rosenb. Malay Arch. p. 554 (1879).—Musschenbr. Dagboek, pp. 212, 241 (1883).—Rosenb. Mitth. orn. Ver. Wien, 1883, p. 54.

Ailuroedus arfakianus, Salvad. Orn. Papuasia, p. 673 (1881).—Sharpe, Cat. B. Brit. Mus. vi. p. 384 (1881).—D'Hamonv. Bull. Soc. Zool. France, 1896, p. 511.—Salvad. Agg. Orn. Papuasia, ii. p. 166 (1890).

THIS is a small form of *Æ. melanotis* of Gray, which inhabits the Aru Islands and the Fly River in South-eastern New Guinea. The Arfak Cat-bird is, however, a slightly smaller bird than *Æ. melanotis*, with a somewhat blacker head and a black band on the nape. The throat also appears blacker than in the Aru bird, by reason of the broader black margins to the feathers.

All the specimens at present known have come from the Arfak Mountains, where they have been found by D'Albertis, Beccari, and the hunters employed by Dr. A. B. Meyer and the late Mr. Bruijn. Count Salvadori had eight specimens before him when he wrote his 'Ornitologia della Papuasia,' and he considers the species to be quite distinct from the Black-cheeked Cat-bird from Aru. Some of the females had the fore-neck less black than in the males, and in all his series the fulvous spots on the crown varied considerably in size, but in one female bird they were especially small and of a reddish-buff colour. A very young female had all the feathers on the lower parts of a soft texture and a brown colour, the upper parts, wings, and tail being of the same colour as in the adult birds.

All references to *Æ. melanotis* from New Guinea, excepting from the Fly River district, doubtless refer to *Æ. arfakianus*.

Professor Schlegel mentions some specimens of a Cat-bird, collected by Hoedt in the Island of Mysol, and refers them to *Æ. melanotis*, but Count Salvadori considers that they are more likely to belong to *Æ. arfakianus* or to an undescribed species.

The Plate represents the specimen procured by D'Albertis in the Arfak Mountains. It is the same one which he lent to the late Mr. Gould for illustration in the 'Birds of New Guinea.'

ÆLURŒDUS MELANOCEPHALUS, *Ramsay.*

Black-naped Cat-bird.

Æluroedus melanocephalus, Ramsay, Proc. Linn. Soc. N. S. W. viii. p. 25 (1883).—Salvad. Ibis, 1884, p. 354.—
Finsch u. Meyer, Zeitschr. ges. Orn. ii. p. 354 (1885).—Iid. Ibis, 1886, p. 238.—D'Hamonv. Bull. Soc.
Zool. France, 1885, p. 511.—Sharpe, in Gould's Birds of New Guinea, i. pl. 42 (1888).—Salvad. Agg.
Orn. Papuasia, ii. p. 166 (1890).—Sharpe, Bull. Brit. Orn. Club, iv. p. xix (1894).—Salvad. Ann.
Mus. Civ. Genov. (?) xvi. p. 114 (1896).

This is the representative of *Æluroedus melanotis* in South-eastern New Guinea, where it inhabits the Owen
Stanley Mountains. It has occurred in Mr. Goldie's collection from the Astrolabe Range, and the late
Carl Hunstein procured specimens in the Horse-shoe Range, while Dr. H. O. Forbes met with it in the
Sogeri District, and Dr. Loria at Moroka.

Æ. melanocephalus belongs to the same section of the genus *Æluroedus* as *Æ. melanotis* and *Æ. arfakianus,*
and is very closely allied to the former, from which it differs in the greater amount of black on the sides of
the face, and in the chin, lores, and the fore part of cheeks being black. The crown of the head is blacker, and
the ovate spots of buff are fewer in number and consequently much more distinct. The breast and abdomen
are of a deeper ochre colour, and the dusky margins to the feathers are much less pronounced. The light spots
on the wing-coverts are apparently variable in extent, and are sometimes absent altogether. Thus Dr. Meyer
described the wing-coverts as uniform, and Count Salvadori has drawn attention to the apparent discrepancy
between Dr. Meyer's statement and the bird figured by me in Gould's 'Birds of New Guinea.' He has,
however, recently found the same differences in specimens sent by Dr. Loria, and in the two examples in
the British Museum the size of the spots on the wing-coverts varies considerably. The fact that my
description and figure given in the above-mentioned work did not entirely correspond, as Count Salvadori
has pointed out, is due to the fact that the specimen figured (from Dr. Forbes's collection) is not the one
described in the text, which is from Hunstein's collection.

Adult male. General colour above grass-green, the upper tail-coverts slightly washed with lighter green;
the upper mantle varied with ovate spots of ochreous buff in the centre of the feathers; wing-coverts like
the back, the median and greater coverts and the bastard-wing faintly tipped with ashy ochreous buff;
primary-coverts and quills externally green like the back, the primaries washed with bluish on the outer web;
the secondaries tipped with ochreous white, less distinct on the primaries; tail-feathers dark green on
the outer web, black internally, all the feathers tipped with white, increasing in extent towards the outer
ones; crown of head black, with ovate spots of ochreous buff, smaller on the forehead and nape, the latter
being almost entirely black; hind-neck ochreous buff, the feathers margined with black; lores black,
surmounted by a line of ochreous-buff-spotted feathers; feathers round eye and ear-coverts black, with a line
of buff-spotted feathers below the eye; behind the ear-coverts a line of whitish down the sides of the neck;
fore part of cheeks black, as well as the chin; throat and sides of neck ochreous buff, mottled with black
edges to the feathers; fore-neck and remainder of under surface of body rufescent ochre, with greenish
edges on the feathers of the chest; the breast and abdomen more uniform; all the feathers with more or less
distinct white shaft-lines; sides of body and flanks like the breast, and washed with greenish; thighs dull
greenish; under tail-coverts like the abdomen, with white shaft-lines; under wing-coverts and axillaries ashy
tipped with whitish; quills below dusky, ashy along the inner edge; "bill greenish white; feet greyish
green; iris hazel" (*L. Loria*). Total length 11·5 inches, culmen 1·3, wing 5·7, tail 4·6, tarsus 1·7.

The figure in the Plate represents an adult male of the natural size, drawn from a specimen collected by
Dr. H. O. Forbes.

ÆLUROEDUS MACULOSUS, Ramsay.

ÆLURŒDUS MACULOSUS, *Ramsay*.

Queensland Cat-bird.

Ælurœdus maculosus, Ramsay, Proc. Zool. Soc. 1874, p. 601.—Sharpe, Cat. Birds Brit. Mus. vi. p. 384 (1881) —
Cairn & Grant, Rec. Austr. Mus. i. p. 31 (1892).—North, l. c. p. 112 (1892).
Ælurœdus maculosus, Gould, Birds of New Guinea, i. pl. 28 (1875).—Ramsay, Proc. Linn. Soc. N. S. W. ii.
p. 187 (1878).—Id. Tab. List Austr. B. p. 11 (1888).—North, Proc. Linn. Soc. N. S. W. (2) ,
p. 1156 (1887).—Id. op. cit. vi. p. 167 (1893).—Id. Nests & Eggs Austr. B. p. 177 (1889).

This species agrees with *Æ. melanotis* and *Æ. arfakianus* in having the crown of the head mottled, the colour in this case being black spotted with olive-brown, while there is also a black patch on the ear-coverts and another on the chin.

Mr. A. J. North has given the following good account of this species in 'Nests and Eggs of Australian Birds' :—

"This bird is a native of the dense scrubs that are to be found in the neighbourhood of Rockingham Bay, and the Johnstone, Russell, and Mulgrave Rivers in Tropical Queensland. They congregate in small flocks on the palms and fig-trees, from which they obtain their food. During a recent excursion to the Bellenden-Ker Ranges, Messrs. E. J. Cairn and Robert Grant, collecting on behalf of the Trustees of the Australian Museum, succeeded in obtaining, amongst others, a fine series of these birds in different stages of plumage ; and, besides finding several nests with young birds, they were fortunate in obtaining, although very late in the season, a nest containing eggs. The nest and eggs in question were found on December 2nd, 1887, in the fork of a sapling about seven feet from the ground, on the Herberton road, at a distance of thirty-two miles from Cairns. The nest is a neat bowl-shaped structure, composed of long twigs and leaves of a *Trietania*, lined inside with twigs and the dried wiry stems of a climbing plant ; on the outside several nearly perfect leaves of the *Trietania* are worked in, and partially obscure one side of the nest. Exterior diameter seven inches, by four inches and a half in depth; interior diameter four inches and three-quarters, by two inches and a half in depth. Eggs two in number for a sitting, nearly true ovals in form, tapering but slightly at one end, of a uniform creamy-white ; the shell is thin, the surface being smooth and slightly glossy. (A) 1·67 × 1·11 ; (B) 1·63 × 1·1 inch. Both parent birds were procured at the time of taking the eggs, which were in a very advanced state of incubation."

Adult. General colour above green ; the wing-coverts like the back, with obscure yellowish spots at the ends of the median and greater coverts ; quills dusky blackish, externally green like the back ; the primaries bluish on their outer webs, the innermost secondaries tipped with a spot of yellowish white ; tail-feathers green, dusky blackish on the inner web, all but the two centre ones barred along the tip with white ; head ochreous brown, mottled all over with blackish edges to the feathers ; the feathers of the hind-neck and mantle greenish, mottled with a spot of pale ochreous or yellowish white ; lores bluish white tinged with yellow ; feathers above and below the eye whitish, the former having blackish tips to some of the feathers, ear-coverts black, with a streak of whity brown tinged with green along the upper edge ; chin and a spot at the base of the cheeks black ; remainder of cheeks yellowish white, the feathers mottled with dusky-brown tips ; sides of neck ashy, tinged with green and mottled with brown edges ; behind the ear-coverts and on the lower neck a whitish shade, forming an indistinct patch ; throat ashy, mottled with greenish-brown edges to the feathers ; remainder of under surface spotted, the feathers having a large ovate mark of white in the centre, and being broadly edged with greenish brown, these margins less distinct on the abdomen, which is consequently whiter ; thighs ashy, washed with green, with a subterminal bar of yellow ; under wing-coverts dusky, tipped with yellowish white forming a faint bar ; axillaries green, with yellowish tips ; edge of wing yellow ; quills dusky below, whitish towards the base of the inner web. Total length 11 inches, culmen 1·3, wing 5·9, tail 4·1, tarsus 1·85.

The description is taken from a specimen in the British Museum, and the Plate represents two figures of an adult bird, together with the nest. For the opportunity of figuring this I am indebted to the kindness of Mr. A. J. North, who sent me an enlarged photograph of it.

ÆLURŒDUS VIRIDIS,(Lath)

ÆLURŒDUS VIRIDIS (*Latham*)

Cat-bird.

Gracu Gouldi, Lath. Gen. Syn. Suppl. ii. p. 129 (1801).
Gracula viridis, Lath. Ind. Orn. Suppl. ii. p. xxvii (1801).—Shaw, Gen. Zool. iii. p. 473 (1809).
Lanius coronirostris, Paykull, Nova Acta Akad. Upsal. vii. p. 282, Taf. 10 (1815).
Kitta viriscens, Temm. Pl. Col. ii. pl. 396 (1826).—Wagler. Syst. Av., *Philædignælus*, sp. 3 (1827).
Philædrhynchus viridis, Vig. & Horsf. Trans. Linn. Soc. xv. p. 264 (1827, ex Latham MSS.).—Gray, Gen. B. ii. p. 325 (1846).—Gould, B. Austr. iv. pl. 11 (1848).
Philædynchus viridis, Bp. Comp. i. p. 350 (1850).—Schl. Mus. Pays-Bas, Coraces, i. p. 117 (1867).
Ailurædus viridis, Cab. Mus. Hein. Th. i. p. 213 (1850).—Gould, Handb. B. Austr. i. p. 446 (1865).
Ælurædus coronirostris, Sclater, Ibis, 1868, p. 301.—Elliot, Monogr. Parad. pl. xxxiv. (1873).—Ramsay, Proc. Linn. Soc. N. S. W. ii. p. 187 (1878).
Philædrhynchus coronirostris, Gray, Hand-l B. ii. p. 204, no. 4336 (1869).
Ælurædus viridis, Sharpe, Cat. Birds in Brit. Mus. vi. p. 385 (1881).—Ramsay, Tab. List Austr. B. p. 11 (1888).—North, Nests & Eggs Austr. B. p. 176 (1889).—Id. Rec. Austr. Mus. i. p. 111 (1891).

In nearly every country in the world there seems to be some bird which has the name of " Cat-bird " bestowed upon it. In North America it is a Mocking-bird, *Galeoscoptes carolinensis*, which bears the name, and in Australia the present species is known to the colonists by the same title on account of the strange resemblance of its note to that of a cat. Structurally and in general appearance the bird is allied to the Bower-birds, but it has never been known to build a " bower." Mr. Gould says:—" While in the district in which this bird is found, my attention was directed to the acquisition of all the information I could obtain respecting its habits, as I considered it very probable that it might construct a bower similar to that of the Satin-bird; but I could not satisfy myself that it does, nor could I discover its nest or the situation in which it breeds; it is doubtless, however, among the branches of the trees of the forest in which it lives." As will be seen below, the nest and eggs are now known, but Mr. North also stated in 1891 that he had never heard of any " bower " being built by the Australian Cat-birds.

The range of the Cat-bird is given by Mr. North as the coastal ranges of New South Wales and Southern Queensland. He writes:—" It is particularly plentiful at Cambewarra and the Kangaroo valley, in the Illawarra district, and is found in favourable localities all through the southern portions of the coast ranges, becoming scarcer, however, as the boundary of the colony is approached. The rich bushes in the neighbourhood of the Clarence, Richmond, and Tweed rivers are also strongholds of this species, and it is also found, but not so freely dispersed, in the extreme south of Queensland." Mr. Gould observes:—" So far as our knowledge extends, this species is only found in New South Wales, where it inhabits the luxuriant forests that extend along the eastern coast between the mountain ranges and the sea; those of Illawarra, the Hunter, the Macleay, and the Clarence, and the cedar-brushes of the Liverpool Range, being, among many others, localities in which it may be found; situations suitable to the habits of the Regent- and Satin-birds are equally adapted to the habits of the Cat-bird, and I have not unfrequently seen them all three feeding together on the same tree."

It should be noted that Dr. Ramsay included Victoria and South Australia among the habitats of the present species; but I presume that these countries were inserted by mistake in the 'Table' of Australian birds, as Mr. North does not include them in his more recently-given range of the species.

Mr. Gould gives the following account of the habits of the Cat-bird:—" The wild fig and the native cherry, when in season, afford an abundant supply. So rarely does it take insects that I do not recollect ever finding any remains in the stomachs of those specimens I dissected. In its disposition it is neither a shy nor a wary bird, little caution being required to approach it, either when feeding or while quietly perched upon the lofty branches of the trees. It is at such times that its loud, harsh, and extraordinary note is heard; a note which differs so much from that of all other birds that, having been once heard, it can never be mistaken. In comparing it to the nightly concerts of the domestic cat, I conceive that I am conveying to my readers a more perfect idea of the note of this species than could be given by pages of description. This concert is performed either by a pair or by several individuals, and nothing more is required than for the hearer to shut his eyes to the neighbouring foliage, to fancy himself surrounded by London Grimalkins of house-top celebrity."

In 1877 Dr. Ramsay described a nest and eggs supposed to belong to the Cat-bird, but he entertained some doubt as to their authenticity, and Mr. North, who reproduced the description in his 'Nests and Eggs of Australian Birds,' evidently shared the misgivings of Dr. Ramsay.

Mr. North writes :—" For an opportunity of examining an authentic nest and egg of the New South Wales Cat-bird, *Ailuroedus viridis*, I am indebted to Mr. W. J. Grime, a most enthusiastic and persevering oologist, who recently procured two nests of this species on the Tweed River, and sent the following notes relative to the taking of them :—

"' On the 4th of October, 1890, I was out looking for nests, accompanied by a boy. I left him for a little while to go further in the scrub, and on my return he informed me he had found a Cat-bird's nest with two eggs, one of which he showed me, the other one he broke when descending from the tree. I went with him to the nest and found the old birds very savage, flying at us, and fluttering along the ground. The nest was built in a three-pronged fork of a tree, about fourteen feet from the ground. The tree was only four inches in diameter, and was in a jungle or light scrub, about fifty yards from the edge of the open country. I felled the tree and secured the nest, of the authenticity of which there is no doubt, as the old birds strongly objected to my taking it. The eggs had been sat on for a few days and were partially incubated.'

"In a subsequent letter dated November the 8th, Mr. Grime writes :—' To-day I found another Cat-bird's nest and drove the parent bird off it myself. I thought I had more eggs, as the Cat-bird would not leave the nest until fairly shaken out, but when I examined the nest I found two young birds in it, apparently just hatched a couple of days.'

"The nest of *Ailuroedus viridis* is a beautiful structure, being bowl-shaped and composed exteriorly of long twigs, entwined around the large broad leaves of *Ptericis aqupropteadron* and other broad-leaved trees, some of the leaves measuring eleven inches in length by four inches in breadth. The leaves appear to have been picked when green, so beautifully do they fit the rounded form of the nest, one side of which is almost hidden by them. The interior of the nest is lined entirely with fine twigs. The nest of *Ailuroedus viridis* is similar to that of *Æ. maculosus*, but larger, and both of them can be really distinguished from those of any other Australian bird by the peculiarity of having large broad leaves used in the construction of the exterior portion of the nest.

"The eggs of *Æ. viridis* are two in number for a sitting, oval in form, being but slightly compressed at the smaller end, of a uniform creamy-white, very faintly tinged with green, the shell being comparatively smooth and slightly glossy. Length 1·66 inch × 1·2 inch."

Adult. Above bright grass-green, with a greyish-white patch on the side of the lower neck ; primaries slightly shaded externally with bluish, the inner webs of the quills greyish brown ; median and greater wing-coverts, as well as the secondaries, tipped with yellowish white ; tail-feathers bright grass-green, inclining to greyish black on the inner web, and tipped with white, more largely on the inner web ; head and neck green, rather more yellowish than the back, the feathers bluish underneath when lifted, the hind neck and sides of the neck, as well as the mantle, slightly streaked with minute shaft-lines of buffy white ; sides of face dull olive-greenish, including the ear-coverts, which have a dull sort of silvery lustre ; round the eye a ring of whitish feathers ; fore part of cheeks, feathers below the eyes and on the molar line, slightly spotted with black ; throat dull greyish, slightly washed with olive-green and minutely spotted with white ; rest of under surface of body dull olive-greenish, all the feathers mesially streaked with a distinct lanceolate spot of white ; centre of the belly, vent, and under tail-coverts uniform yellowish ; under wing-coverts whitish, barred across with ashy grey and slightly washed with green, especially on the edge of the wing ; " bill light horn-colour ; feet whitish ; iris brownish red" (*J. Gould*). Total length 12 inches, culmen 1·25, wing 6·35, tail 5, tarsus 1·95.

The description is taken from the British Museum 'Catalogue of Birds,' and the figures in the Plate represent a pair of adult birds, drawn from specimens in the British Museum.

ÆLUREDUS BUCCOIDES (Temm).

ÆLURŒDUS BUCCOIDES (*Temm.*).

Barbet-like Cat-bird.

Kitta buccoides, Temm. Pl. Col. ii. pl. 575 (1835).—Id. Tabl. Méth. Pl. Col. i. p. 10 (1840).—Rosenb. Nat. Tijdschr. Nederl. Ind. xxv. p. 236 (1863).—Id. J. f. O. 1864. p. 122.

Crasa buccoides, Gray, Gen. B. iii., App. p. 14 (1849).

Ptilorhynchus buccoides, Bp. Consp. Av. i. p. 350 (1850).—Wallace, Proc. Zool. Soc. 1862, p. 165.—Id. Ann. & Mag. N. H. (3) xi. p. 37 (1863).—Schl. Mus. Pays-Bas, Coraces, p. 118 (1867).—Id. Nederl. Tijdschr. Dierk. iv. p. 49 (1871).

Ptilonorhynchus buccoides, Gray, Proc. Zool. Soc. 1858, p. 194.—Id. Cat. B. New Guin. pp. 37, 50 (1859).—Finsch, Neu-Guinea, p. 173 (1865).—Gray, Hand-l. B. i. p. 294, no. 4337 (1869).—Rosenb. Malay. Archip. p. 554 (1879).—Meyenborg. Dagboek, pp. 231, 240 (1883).—Rosenb. Mitth. orn. Ver. Wien, 1885, p. 56.

Ælurœdus buccoides, Gould, B. New Guinea, i. pl. 41 (1875).—Salvad. Ann. Mus. Genov. vii. p. 760 (1875).

Ælurœdus buccoides, Elliot, Monogr. Parad. pl. 36 (1873).—Scl. Proc. Zool. Soc. 1873, p. 697.—Salvad. Ann. Mus. Genov. ix. p. 193 (1876), x. p. 152 (1880).—D'Alb. et Salvad. op. cit. xiv. p. 114 (1879).—D'Alb. Nuova Guin. pp. 581, 588 (1880).—Salvad. Orn. Papuasia, ii. p. 675 (1881).—Sharpe, Cat. B. Brit. Mus. vi. p. 386 (1881).—Nehrk. J. f. O. 1885, p. 34.—Guillem. Proc. Zool. Soc. 1885, p. 657.—Salvad. Agg. Orn. Papuasia, ii. p. 167 (1891).—Madarász, Aquila, i. p. 91 (1894).

This is one of the smaller species of Cat-bird, and is further distinguished by the strongly marked spotting of the breast and the very distinct white streaking on the nape, and by the absence of white tips to the tail-feathers.

The first specimen was procured by Solomon Müller at Lobo, in Triton Bay in New Guinea, and it seems to be somewhat widely distributed in that great island, for Count Salvadori gives the following localities from which he has examined specimens—Sorong, Dorey, Mansinam, Andai, and Warbusi. The species was obtained in these places by Dr. Beccari, Signor D'Albertis, and by the hunters employed by the late Mr. Bruijn. D'Albertis also met with this Cat-bird on the Fly River, and the late Mr. Fenichel likewise procured a specimen during his expedition to the Finisterre Mountains, in German New Guinea, at a place called Kulikomana, on the 29th of August, 1892. Dr. von Madarász was so kind as to send me this specimen for examination. It seemed to differ slightly in the colour of the head from our series of skins of *Æ. buccoides* in the British Museum; but after the remarks of Count Salvadori on the variation in the colour of the crown in the present species, I could not regard it as belonging to anything else. Certainly it was not *Æ. geislerorum*, of which specimens were also sent by Fenichel.

Besides the above-mentioned places in New Guinea, examples of *Æ. buccoides* are in the Leiden Museum from the islands of Salawati, Waigiou, and Batanta, where they were procured by the late Dr. Bernstein. Von Rosenberg's statement that it is also found in the Aru Islands must be erroneous, as already pointed out by Count Salvadori.

Nothing has yet been recorded concerning the habits of this Cat-bird, beyond the fact that D'Albertis found it feeding on fruits.

The following description is a copy of that given by me in the sixth volume of the 'Catalogue of Birds':—

Adult. Above bright grass-green, the wings uniform with the back; primaries blackish, externally bright green, the secondaries slightly shaded with bluish on the outer web, the innermost minutely tipped with yellowish buff; tail duller green, narrowly tipped with white on the inner web of the outer feathers; crown of head olive-brown, the hinder neck, as well as the sides of the latter, black, streaked with yellowish buff, this colour occupying the basal part of the feather, the black confined to a large subterminal spot; the mantle also slightly mottled with yellowish buff, with which many of the feathers are barred, some few being also tipped with black; lores scantily feathered with brown plumelets; eye-ring buff; behind the eye a bare space; sides of face buff, everywhere mottled with black spots, the hinder part of the ear-coverts entirely black, the cheeks also somewhat spotted with white; throat buffy white, the chin and lower throat spotted with black; rest of under surface light fawn-buff, everywhere largely spotted with ovate black markings, these spots less on the abdomen and absent on the under wing- and tail-coverts, which are uniform; the breast, flanks, and thighs obscurely washed with green, some of the spots on the lower flanks being green also. Total length 10 inches, culmen 1, wing 4·85, tail 3·5, tarsus 1·35.

Count Salvadori states that the females are smaller than the males. The crown of the head is uniform brownish olive, but in some individuals it inclines more or less to olive or to brown. Young birds have the head of an olive-brown colour, more or less clear in tint.

The figure represents a specimen formerly in the Gould collection, and now in the British Museum. It is of about the natural size.

ÆLUREDUS GEISLERORUM, Meyer.

W.Hart del. et lith.

Mintern Bros. imp.

ÆLURŒDUS GEISLERORUM, *Meyer*.

Geisler's Cat-bird.

Ælurœdus geislerorum, Meyer, Abhandl. k. zool. Mus. Dresden, 1890-91, no. 4, p. 12.—Id. J. f. O. 1892, p. 297.—Id. Abhandl. k. zool. Mus. Dresden, 1892-93, no. 3, p. 23 (1893).—Madarász, Aquila, i. p. 91 (1894).—Sharpe, Bull. Brit. Orn. Club, iv. p. xiv (1894).

This Cat-bird was described by Dr. A. B. Meyer from specimens obtained in Kaiser Wilhelm's Land in German New Guinea by the brothers Geisler, after whom the species has been named. The original specimens were procured in Astrolabe Bay, at Lolobu and Bussum, to the north of Huon Gulf, in July, August, December, and January; and the species was also obtained by the Geislers at a place called Butarieng, which is one of the stations of the German New Guinea Company, and is situated near the mouth of the Babui River. Other localities in which the species was met with are—Finsch Harbour in March, Bukawanip in Huon Gulf in April, Mensuing in June, and again at Butarieng in August and October. The brave Hungarian traveller, Fenichel, who lost his life in exploring this portion of New Guinea, also obtained specimens of Geisler's Cat-bird at Bonga in November. And this apparently is all that is known respecting the range of *Ælurœdus geislerorum*.

The species is allied to *Ælurœdus buccoides*, but has a very differently coloured head, this being much paler and more of a tawny-olive shade, while the spots on the breast are much larger and more pronounced.

Little has been recorded of its habits. Fenichel says that its note resembles the sound "Aach-aach," and it is called by the natives "Kubuss." The Geislers state that the species is generally distributed in the parts of the country visited by them, but could not be considered common. It is always found in pairs. The male suddenly utters his note of warning, a long-drawn, hoarse-sounding "tschää," the female quickly answering and repeating the call. In this way the bird is continually in evidence to the collector, and is easy to kill, though it requires a practised eye to detect the bird amongst the surrounding foliage.

The following is the description of a specimen collected by the late Mr. Fenichel :—

General colour above grass-green, with the wings of the same colour as the back, the inner secondaries with a whitish tip to their outer webs; bastard-wing, primary-coverts, and quills blackish, externally green, inclining to olive towards the end of the primaries; tail-feathers blackish, externally green, inclining to ashy white at the extreme tips; crown of head tawny olive or ochreous brown; lores and sides of face and ear-coverts white; chin and cheeks spotted with black, forming a broad moustachial band; throat white; sides of neck entirely black, forming a large patch behind the ear-coverts and crossing the nape in a black band; hinder neck pale yellowish, washed with green, and having large triangular spots of black; remainder of under surface of body, from the lower throat downward, pale ochreous yellow, with a slight tinge of greenish on the flanks, the breast and sides of the body thickly marked with large triangular spots of black, becoming smaller on the lower part of the abdomen and disappearing entirely on the under tail-coverts; under wing-coverts white; axillaries ochreous buff; bill pale yellowish white; iris carmine. Total length 10·5 inches, culmen 1·05, wing 5·2, tail 3·65, tarsus 1·45.

Dr. Meyer mentions that two specimens collected by the brothers Geisler at Bukawanip had the iris brown.

The Plate represents an adult bird in two positions, and is drawn from the specimen above described.

ALCIPPE'S STONII, *Sharpe*

ÆLURŒDUS STONII, *Sharpe.*

Stone's Cat-bird.

Ælurœdus stonii, Sharpe, Nature, xiv. 1876, p. 339.—Salvad. Ann. Mus. Gen. ix. p. 193 (1876).—Sharpe, Journ. Linn.
   Soc. xiii. p. 495 (1877).—Ramsay, Proc. Linn. Soc. N.S.W. ii. p. 269 (1879); iv. p. 97.—Salvad. Orn.
   della Papuasia, ii. p. 678 (1881).—Gould, B. New Guinea, i. pl. 37 (1881).—Sharpe, Cat. B. Brit. Mus.
   vol. vi. p. 387 (1881).—Id. Journ. Linn. Soc. xvi. p. 445 (1883).—Finsch u. Meyer, Zeitschr. ges. Orn. ii.
   p. 391 (1885).—Id. Ibis, 1886, p. 238.—D'Hamonv. Bull. Soc. Zool. France, xi. p. 414 (1886).—Salvad.
   Agg. Orn. Papuasia, ii. p. 167 (1890).—De Vis, Rep. Brit. New Guinea, 1890, p. 116.
Ælurodus stonii, Ramsay, Proc. Linn. Soc. N.S.W. iii. p. 268 (1879).—Id. op. cit. iv. p. 97 (1880).
Ptilonorhynchus stonii, Van Musschenbr. Dagboek, pp. 212, 241 (1883).—Rosenb. Mitth. orn. Ver. Wien, 1885,
   p. 54.

THE present species is one of the smallest of the Cat-birds, and seems to be entirely confined to the south-eastern portion of New Guinea, where it replaces *Ælurœdus buccoides* of North-western New Guinea. It is a smaller bird than the latter, and further differs in having the crown of the head blackish brown instead of olive-brown, and in having the spots on the throat and under surface of the body very much smaller.

It was first met with in South-eastern New Guinea by Mr. Octavius Stone, who found it on the Laloki River, and its home, so far as is known at present, appears to be the interior of the Port Moresby district. Mr. Goldie found the species about fifteen miles inland, inhabiting the dense scrub and feeding on fruits and berries. He afterwards procured it in the Sogeri district, where it was called by the natives " Yaritaggo." Mr. Forbes likewise met with it in Sogeri. The late Karl Hunstein procured the species on the Horse-shoe Range, and Sir William Macgregor also obtained specimens on Mount Belford, in the Astrolabe Range, at 4000 feet.

Mr. Goldie forwarded two eggs supposed to be those of *Ælurœdus stonii* from Sogeri, but as they were white, and entirely different from those of the ordinary Cat-birds, it is reasonable to suppose that the identification was not correct.

The following is the description of the type specimen given by me in the 'Catalogue of Birds':—

Adult. General colour above bright green, some of the feathers tinged with blue ; wings green like the back, the inner webs dusky brown, the primaries externally washed with yellow, the secondaries tipped with the latter colour ; tail green, blackish on the inner webs of the outermost rectrices, which are tipped with white ; head dark brown, slightly washed with olive ; hind neck yellowish buff, mottled with black centres to the feathers, those adjoining the mantle spotted with green ; sides of face and throat pure white, with a few tiny spots of black on the ear-coverts, and with larger spots on the sides of the neck ; rest of under surface of body ochraceous buff, the fore neck and chest minutely spotted with green, the flanks also with a few tiny spots of the latter colour ; under wing-coverts yellowish buff, the edge of the wing washed with green. Total length 9·3 inches, culmen 1·15, wing 5·05, tail 3·5, tarsus 1·55.

The Plate represents an adult bird of the natural size.

ÆLURŒDUS JOBIENSIS, Rothsch.

Jobi-Island Cat-bird.

Æluroedus jobiensis, Rothschild, Bull. Brit. Orn. Club, iv. p. xxvi (1895).

This species has been described by the Hon. Walter Rothschild from a specimen collected in the Island of Jobi by the late Mr. Bruijn's hunters. I have examined the type-specimen in Mr. Rothschild's museum, and at first sight it looks very distinct from *Æ. arfakianus*, and has the hinder neck much mottled with buff, with the blackish shade overspreading the throat; but there are other specimens in the Rothschild museum which appear to me to be intermediate.

The following is Mr. Rothschild's description:—" This species is nearest to *Æ. melanocephalus* of Ramsay, from British New Guinea, but shows sufficient differences to justify its separation. The head is black, uniformly spotted with buffish yellow, and does not show the black band on the sides of the occiput so conspicuous in *Æ. melanocephalus*. Upper neck and back brownish buff, with black margins. Ear-coverts consisting of the large patch of bristly feathers found in its three nearest allies; but this patch passes straight into the black of the throat, without any marked area of pale feathers surrounding it, as in *Æ. melanotis*, *Æ. melanocephalus*, and *Æ. arfakianus*. The pale spots on the tips of the wing-coverts not very distinct and of a dusky buff colour. Throat, breast, and apperrmost part of abdomen black, with a small central buff spot in each feather, while in *Æ. melanotis* (from the Aru Islands) and *Æ. melanocephalus* these feathers are buff or whitish, with narrow black borders. The breast is much greener in *Æ. arfakianus* from Mt. Arfak. Lower abdomen and under tail-coverts buff with dusky margins, shaded here and there with green. In all other respects most similar to *Æ. melanocephalus*, but the feathers on the sides of the neck just behind the ear-coverts are almost uniform buff, having nearly lost their dark margins. Culmen 1·5 inch, wing 6·5, tail 5·4, tarsus 1·65."

I have not considered it necessary to give a figure of this species.

.

TECTONORNIS DENTIROSTRIS *(Homeny)*

TECTONORNIS DENTIROSTRIS, *Ramsay.*

Tooth-billed Bower-bird.

Scenopæus dentirostris, Ramsay, P. Z. S. 1875, p. 591.—Ed. Proc. Linn. Soc. N. S. W. ii. p. 189 (1878).—Gould,
B. New Guinea, i. pl. 43 (1880).—Sharpe, Cat. B. Brit. Mus. vi. p. 394 (1881).—North, Proc. Linn.
Soc. N. S. W. (2) i. p. 1162 (1887).—Ramsay, Tab. List Austr. B. p. 11 (1888).—De Vis, Rep. Exped.
Bellenden-Ker Range, p. 86 (1889).

THIS curious Bower-bird was first described by Dr. E. Pierson Ramsay from specimens shot by Inspector
Johnstone near Cardwell in Queensland. Unfortunately they were killed with a rifle and the skins were
somewhat damaged, but were sufficiently intact to allow of their correct description; and one of them was
sent by Mr. Coxen to the late Mr. Gould, who figured it in his 'Birds of New Guinea.'

The late Mr. Bowyer Bower obtained several specimens during his first expedition to North-eastern
Australia, and a pair were presented by him to the British Museum.

With the first discovery of the species came but little information regarding its habits, but Inspector
Johnstone stated that, although he did not discover a 'bower,' he found that, like the Cat-birds (*Ælurœdus*),
it cleared a large space under the brushwood some nine or ten feet in diameter and ornamented the cleared
part with tufts and little heaps of tinted leaves and young shoots. This characteristic habit, which places
Tectonornis among the "Gardener" Bower-birds, is confirmed by that excellent observer, Mr. A. J. North,
who writes:—

"This remarkable bird is quite unlike any other genus of the family, and is found only in the dense
brushes of the Bellenden-Ker Range, situated on the north-east coast of Queensland; its range does not
extend further north than the scrubs near Cooktown, nor has it been found further south than the Herbert
River. As far as at present known, this species does not build a bower; but in lieu thereof clears a space
in the scrub about 10 feet in diameter, and ornaments it with little heaps of bright berries and gaily
coloured leaves and flowers, &c. Nothing is known of its nidification at present."

Mr. Broadbent, who also met with the species in the Bellenden-Ker Range, says that it is a true mountain
bird and was procured at 4000 feet : it was not found in low scrubs, or at least very seldom.

The name of *Scenopæus* being already preoccupied among the Diptera, I have proposed that of *Tectonornis*
for the present species.

Adult male. General colour above dark olive-brown, wings a little more ruddy brown than the back, quills
dusky on the inner web; tail olive-brown like the back; ear-coverts dusky brown, streaked with mesial
shaft-lines of reddish brown, the cheeks and sides of the neck similarly streaked with broader pale streaks;
under surface of body fulvous, the feathers with broad margins of dusky brown, imparting a streaked
appearance, less strongly marked on the abdomen; under wing-coverts and axillaries deeper fulvous, of a
tawny shade, slightly mottled with a few dusky edgings to the feathers; quills dark brown below, yellowish
along the edge of the inner web. Total length 10·5 inches, culmen 1·15, wing 5·7, tail 3·7, tarsus 1·3.

Adult female. Similar to the male in colour. Total length 10·5 inches, culmen 1·2, wing 5·6, tail 3·45,
tarsus 1·35.

The figure in the Plate is taken from the type specimen lent by Dr. Ramsay to the late Mr. Gould.
The descriptions are taken from examples in the British Museum.

www.ingramcontent.com/pod-product-compliance
Lightning Source LLC
Chambersburg PA
CBHW021711210326
41599CB00013B/1618